THE FULLY CHARGED GUIDE TO

Electric Vehicles & Clean Energy

THE FULLY CHARGED GUIDE TO

Electric Vehicles & Clean Energy

ROBERT LLEWELLYN and his *Fully Charged* friends

unbound

First published in 2020

Unbound
6th Floor Mutual House, 70 Conduit Street, London W1S 2GF
www.unbound.com
All rights reserved

Designed for Essential Works Ltd. (essentialworks.co.uk)
by seagulls.net

A CIP record for this book is available from the British Library

ISBN 978-1-78352-858-5 (hardback)
ISBN 978-1-78352-859-2 (ebook)

Printed in [xxx] by [xxx]

1 3 5 7 9 8 6 4 2

Fully Charged, founded and hosted by Robert Llewellyn (*Red Dwarf*, *Scrapheap Challenge*, *Carpool*), is the world's number-one clean energy and electric vehicle YouTube channel. It's not only about electric cars, bikes, boats and planes but how we generate and even own the electricity to power these machines. From looking behind the myths of renewable energy to seeking the truth about conventional generation, Robert and his *Fully Charged* friends reveal what the future of energy and transport has stored up.

Find out more at FullyCharged.Show

FULLY CHARGED

CONTENTS

THE FUTURE IS ALREADY HERE, HERE & HERE

ENERGY & TRANSPORT FOR ENQUIRING MINDS

NEXT STEPS FOR YOUR JOURNEY

ABOUT *FULLY CHARGED*

INTRODUCTION

*Robert gets ready to take the renewable-powered
Nissan LEAF on a tour of Tenerife – it's a tough job*

THE BIG SWITCH

Electric cars: the transition from ICE to NICE

Robert Llewellyn

The one thing I've learned in the last decade is not to panic. No one knows what the future will bring, something I was reminded of the other day when I read that Isaac Asimov, writing in 1984, predicted that we'd all be zipping around in fully autonomous cars by 2019 and human beings would be banned from driving. Not quite yet, Isaac, but good guess, mate.

What we *can* say is we are seeing a very big change around the world; we don't need to predict that, it's clearly already happening. I won't drown you in statistics and figures – there's plenty of those in the rest of this book – but every year the amount of renewables installed, the number of electric cars sold and the number of electric

buses and taxis taking to our cities' streets is growing almost vertically. The development of new systems linking these technologies together is also experiencing enormous growth.

When I started making the *Fully Charged* show back in 2010, I hoped – not predicted, *hoped* – that electric cars, renewable energy and all the associated technologies that I was learning about would develop into viable alternatives to the current unsustainable system we live in.

I had very little to base this hope on. I was always prepared to wake up one morning and discover that the whole thing was a pipe dream and once again I'd been an idealistic fool.

Back then electric cars were – as every motoring journalist,

'petrolhead' and anyone with an opinion about it was more than happy to tell me – a total joke.

The future, I was told again and again, was hydrogen and nuclear. It wasn't suggested as a possible alternative, or something to go along with battery technology and renewable energy sources, and it was never presented as an argument; it was simply stated as a fact. Hydrogen is the future. Nuclear is clean, safe and also the future.

So when battery electric cars started to make a comeback, the reaction against this development was loud, slightly angry and powerful. That reaction has diminished and the levels of interest and understanding have increased.

1. Electric buses charging in Waterloo bus garage, London, UK (2017)
2. Linkker buses, the first fast-charging electric buses in Finland (2018)
3. Electric buses parked in a lot in Chongqing, China (2018)
4. Fully electric Heuliez city bus with solar panels in Velika Gorica, Croatia (2018)
5. A transit system village rider bus in Davidson, North Carolina, USA (2019)

1. Teo Taxi, Montreal's first fully electric taxi service, available to anyone in Montreal with a smartphone (2017)
2. Electric taxi charging station in Edinburgh city centre (2019)
3. Black cab charging on the street with fast charge in Coventry, UK (2018)
4. Electric taxi plugged into a charge station in Warsaw, Poland (2017)
5. A fleet of Tesla luxury electric taxis at Amsterdam Airport Schiphol (2015)

As with all new technologies, the new ideas and connections, previously unimaginable, emerge out of the mist as new technologies become abundant. New technologies disrupt the existing status quo, causing upset and a strong push back, but if they are economically viable, they always win.

This is, I now firmly believe, what we're seeing.

Every prediction since 2010 about falling costs, rising demand and a shift in public perception is being borne out by a constant torrent of reports and financial briefings. Sustainable technology, the best all-embracing term I've heard to describe what *Fully Charged* is about, is getting cheaper, more effective and longer term, and will cause the 'big switch' within the lifetime of anyone reading this.

This switch from burning finite fuels to using the power of the sun (which includes wind turbines) is the very essence of disruptive technology, and I've always been fascinated by the effect of disruptive technology.

I just want to state right at the start that disruptive technology is never *all* good. I have lived through the disruption caused by the digital revolution, and I never saw that disruption coming – it took me by surprise. Mobile phones and the internet are both things I use, but I'll be honest, I had no idea at any point during the development of these technologies just how massive their impact would be on the world and how fast it would happen. The digital revolution had a profound impact on the TV industry that I've been working in for the last 30 years. It affected me personally, particularly my income, because people used to buy DVDs. Remember those?

So we are set to see big changes, not just with electric cars, but with the electrification of virtually all transport. Of course hydrogen has a huge part to play in that – possibly not in passenger cars, but ships, trains, trucks and heavy machinery are all prime candidates for hydrogen technology. And just to illuminate that a little, the resulting drive system will be electric motors, but the energy storage might be hydrogen and fuel cells.

The one thing I really didn't imagine 10 years ago is electric aeroplanes. I never even thought about them but clearly that is one area we are going to see electrified much sooner than expected. EasyJet, of all companies, are investing billions in electric plane development.

So, apart from all the political upheaval, the rise of nationalism, extremism, fundamentalism and all the other 'isms', we are living through very exciting times. I hope we have captured the essence of that within these pages, and I hope you enjoy reading about them and glean some optimism from our efforts.

Remember not to panic next time you see an article with the headline 'Electric Cars Will Literally End the World'.

> *We are set to see big changes, not just with electric cars, but with the electrification of virtually all transport*

THE LIGHTNING RODS

From Nikola Tesla to Elon Musk

Ben Sullins

Nikola Tesla was a stud. Born in 1856 in what is today Croatia to an Eastern Orthodox priest and a mother with a knack for inventions, he would grow up to become a famous engineer and inventor in the United States. In high school, Tesla became fascinated with demonstrations of 'mysterious phenomena' when experiencing displays of electrical power in his physics classes. With an eidetic memory he credited to his mother, Tesla excelled at engineering and maths. After high school, Tesla returned home only to contract cholera and nearly die several times that summer. During this time his father promised to send him to the best engineering school if he survived. That, of course, turned out to be a great decision for Tesla, and for the rest of us.

Tesla studied mechanical and electrical engineering at the technical university in Graz, Austria, knowledge that eventually landed him a job in Paris for the Continental Edison Company. His outstanding work there impressed some of the bosses, which led them to offer him a position in the United States.

In America, Tesla initially struggled working for Thomas Edison's company as an inspector of their new indoor lighting systems. But his time there was well spent learning about American business and commercial practices, which differed greatly from those in Europe.

When Tesla struck out on his own, he invented many electrical-based motors and systems, including the alternating current (AC) induction motor. This new type of motor had major advantages over other available options at the time, including a self-starting design that reduced maintenance and the need to replace mechanical brushes.

His new venture selling these motors went well, earning Tesla large sums of money and a royalty deal on every motor sold. It wouldn't last forever, though. Later, after the AC motor money had dried up, Tesla continued to invent fascinating new electrical machines including wireless electric power. He demonstrated this by lighting

Nikola Tesla, c.1895

incandescent bulbs from across a stage wirelessly. Yes, Tesla invented wireless electricity in 1890. Stud.

Fast-forward to Tesla Motors in 2019: they have continued to push the boundaries of what electric vehicles (EVs) can be, benefiting greatly from the early work of the company's namesake. More on this later.

The challenge with EVs has long been the batteries. Well, that and how to actually convert that stored energy into motion. These happen to be two things that J. B. Straubel, Tesla's chief technology officer, was experimenting with in the early 2000s when he met Elon Musk.

Musk, who had recently sold his stake in PayPal for around $165 million, was interested in this idea, yet unsure of where it was going at the time. But, before we continue with Musk building up Tesla into what it is today, we need to understand how Musk and Tesla were similar.

They both emigrated to the United States with backgrounds in physics and engineering. Musk grew up in South Africa with his mother Maye and siblings Kimbal and Tosca. As he grew up he decided to leave his mother's home and move in with his father Errol. This turned out to be a bad idea.

Musk was a nerdy kid in high school who regularly received beatings from other kids. His father apparently had no sympathy for him either, pushing him into a constant state of suffering during his childhood.

The silver lining from this time in Musk's life appears to be his relentless pursuit of making the world a better place, along with his unmatched work ethic. Many have criticised him for his obsessive dedication to his work at Tesla, SpaceX and other projects. In a similar fashion, Nikola Tesla endured some rough times as a child but ultimately made his way to the United States to become an entrepreneur and inventor who would alter human history.

So when Musk was presented with the opportunity to invest in a budding start-up called Tesla Motors in 2003, he jumped at the chance. The company, founded in July that year by engineers

> They have continued to push the boundaries of what electric vehicles (EVs) can be

Elon Musk at the worldwide debut of the Tesla Model X in Los Angeles, California (2012)

Martin Eberhard and Marc Tarpenning, was aiming to produce electric vehicles for the masses.

The strategy was first to build a sports car, the Tesla Roadster, and prove that electric cars could be sexy and fast. They succeeded at this. Musk was deeply involved with the design and engineering of the original Roadster, which was based on the Lotus Elise body but modified to fit Tesla's motors and battery pack.

A key innovation here was a throwback to Nikola Tesla's invention of the AC induction motor. At the time of writing, this design continues to be the most powerful and high-performing electric motor used in an automobile.

Not long after the company began making the Roadster, the founders left the company, making Musk the chief executive officer and chairman of the board. This was in addition to J. B. Straubel, who assumed his role as chief technical officer.

The original Tesla Roadster did what it was intended to do: capture the world's imagination and open up people's eyes to the possibility of electric vehicles they would actually want to own and drive. From there, Tesla, guided by Musk, continued on to the second part of their master plan: a luxury sedan.

Codenamed WhiteStar, Tesla's five-seat luxury sedan was unveiled as the Model S on 30 June 2008. Designed by newly hired Franz von Holzhausen after a failed attempt by Henrik Fisker, the Model S drew international attention similar to the Roadster. While still an expensive vehicle, the Model S was more affordable than the Roadster and far more practical.

Full-scale production and deliveries of the Model S didn't come for a few more years, however. After its commercial launch in 2013 it won numerous awards including being named the 'Car of the Century' by *Car and Driver* magazine. It is no exaggeration to say the Model S put Tesla on the world stage and had a major impact on the world's perception of the possibility of owning an EV.

Of course, the Model S was (and is) out of reach for most people. This is where Elon Musk took another huge risk: as he stated, 'the reason Tesla exists' is to design and produce a mass-market vehicle. Codenamed BlueStar, the Tesla Model 3 is a more affordable version of the Model S sedan, with stripped-down features meant to offer the same driving experience but for much less money.

The Model 3 nearly broke Musk and Tesla. In an interview

on the US TV news programme *60 Minutes*, Musk stated the company was within 'single-digit weeks' of death.

This wasn't the first time Tesla nearly died. In 2008 the company was out of cash, and with the world in a global financial crisis no one was standing in line to invest. When Musk and the team were able to develop a battery pack for Daimler's Smart car, this last-minute push secured a contract with Daimler as well as a $50 million investment.

Since the Model 3 'production hell' has wound to a close and volume production of the car has stabilised, Musk has stated that this was the last 'bet the company' endeavour they would take on. For those of us who follow Tesla and Musk, however, we have our doubts.

Since Tesla began delivering the Model 3 at scale in early 2018 in the United States, it has continued to climb the sales charts, finishing in the top five passenger cars sold in the USA for the last three months of 2018. No other car has ever come close to competing with regular combustion engine car sales until this point, and the trend continues to rise, pushing Tesla ever closer to being the bestselling car brand in the USA. They are, of course, the only brand to focus on

pure battery electric vehicles and their success hasn't gone unnoticed.

Due to Musk's relentless pursuit of progress and constant innovation, Tesla have forced the hand of all major automakers to invest in electric vehicles. Ford Motors in the USA have committed to an $11 billion investment in EVs, while Audi in Germany have committed €14 billion. These are just two examples of major automakers playing catch-up with Musk and Tesla in the EV space.

The pace of electric vehicle adoption is accelerating almost as quickly as the cars themselves. It seems every month a new EV start-up hits the scene with a fancy new model that expands on the existing electric offerings. Sometimes these fall flat, but in any event, they continue to push for what Musk has wanted all along: a switch to sustainable forms of transport.

Musk is so committed to this mission that he stated in his interview with *60 Minutes* that if another company made far better electric vehicles and forced Tesla into bankruptcy, he would still consider it a success. Of course, he wouldn't sleep on that for long and would likely start a new company to try to change the world for the better,

just as Nikola Tesla did after his AC motor company failed.

Elon Musk and Nikola Tesla: two different centuries, two different worlds, but two men driven to advance the human race with technology.

Tesla cars at Tesla Supercharger stations near Atlanta, Georgia (2019)

Due to Musk's relentless pursuit of progress and constant innovation, Tesla have forced the hand of all major automakers to invest in electric vehicles

DISRUPTIVE INFLUENCES

Disruption and convergence – the perfect storm

David Hunt

The future of energy and transport will not be an iteration of what we have now – the game is truly changing. In the mainstream media you may see talk of a gradual move towards electric vehicles (EVs) by 2050. You'll hear how renewables will continue to increase, you'll start to see a handful of large-scale batteries integrated into electricity grids, but for the foreseeable future, it's all too expensive. Iterative thinking. But then, that is how the 'experts' and management consultants have been taught to think; it's how things have often been. Let's face it, the internal-combustion engine (ICE) car has evolved over the last 100 years, and pretty much in tandem, the oil, gas and electricity systems and their associated businesses have evolved over the same period. Sure, the experts say, technology might speed up a bit, but most forecast significant change by 2050. Wrong.

We are at a time when a multitude of technologies and business models are appearing, accelerating and converging at such a pace that significant disruption occurs. Let's take a recent example: the smartphone. Around 12 years ago the smartphone didn't exist – not until Apple launched their first iPhone in June 2007. Who now can live without theirs? Who still has a camera, or buys maps, encyclopedias or newspapers? The list goes on. I use mine to communicate (email, phone, messaging, social media), consume news, read books, magazines and articles, watch TV, shop, pay bills, transfer money, book travel, find my way around, Google anything I want to know or see, and in particular, of course, watch funny videos of cats.

The smartphone could not have happened even one or two years earlier because it took advances in several technologies to converge, namely the reduced cost of computing power, the increased energy density and reduced cost of lithium batteries, the increased capacity and reduced cost of data storage, and the advent of 'cloud' computing. Without any of these being at the right capability, and the right cost, the smartphone would not have developed when and how it did. Business-model innovation developed too: I pay a monthly fee, and with this I fund the purchase of the phone and can send and receive unlimited calls and texts, download and use unlimited apps, and use as much data as I need. More on this business model later.

Disruptions happen quickly when things converge. The smartphone killed the 'traditional' mobile phone in less than 10 years. And who still uses a landline? The digital camera killed the photographic film industry in less than 10 years too. Kodak dominated that market and went bust, just as Nokia dominated the original mobile phone market and went bust.

Energy disruption
Disruptions escalate at different times in different markets, based on local conditions, but they still happen very quickly and globally. Renewable technologies have been the cheapest form of electricity generation in some parts of the world for some time, usually in developing countries, but now, in most of the developed world solar and wind are the cheapest forms of power generation, and this is having the expected effect. Utilities and traditional power generation companies are being disrupted and are going bust. At the time of writing, nuclear power projects are being cancelled around the world, including in Europe and

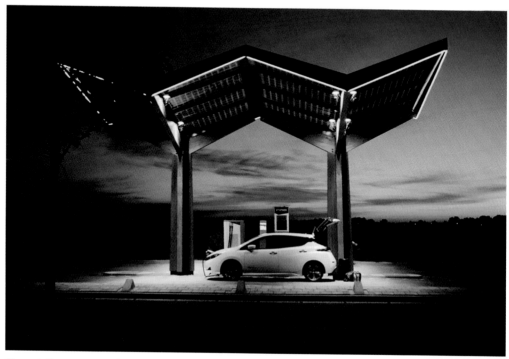

New solutions are emerging at the intersection of energy and transport

the USA, and leading nuclear contractors are going out of business. It's not just nuclear; utility companies are failing and being replaced by new business models, while others are disposing of their fossil fuel assets and going full on for renewables.

Again, we have convergence. The cost curve of deploying renewables has decreased rapidly and shows no sign of stopping. In 2007, when I got into the solar sector in the UK, the cost of a watt-peak of solar was £2.83; by the start of 2019 it was about 35p. And of course, batteries (energy storage) have been the big game changer. Batteries have benefited from their own significant reductions in cost and increases in energy density, with both showing no sign of slowing. But batteries don't just shift power from when it's generated to when it's needed, overcoming the 'intermittency' of solar and wind. They also provide a host of other services depending on location and local energy market rules, such as grid balancing, frequency response, capacity reserve and arbitrage – the practice of charging during lower-priced hours and discharging during higher-priced hours – among others.

Transport disruption

As I mentioned, 'experts' think in a linear fashion, a case of replacing ICE cars with EVs on a like for like basis. That is not going to be the case. I think we have passed 'peak car'. Yes, there will be a fast transition from ICE to EV – I can't see many new cars not being fully electric or at least hybrid by 2025 (the vast majority will be full EV). The cost of cars, falling largely because of the rapid drop in battery costs, combined with the total cost of ownership (TCO) – fuel/charging, maintenance, etc. – will mean it's a simple economic choice to go EV, not a moral or environmental one. Electric vehicles are just better than ICE ones by every measure (unless you love noise). The bigger transition will be the fleet market and light commercial, where the cost and TCO are even more important factors.

Of course, it's not just cars and vans but trucks, tractors, ships, ferries and planes. All will be disrupted by batteries in the very near term. And we haven't even started on autonomous driving – what a revolution that will be. Again, that convergence of increased capability and reduced cost in cloud computing, data storage, computing power, 4G/5G and

and business models that caused the speed and scale of these disruptions.

The future will be 'energy as a service' and 'mobility as a service', or 'transport as a service', as it's referred to in the USA. And both of these things are already happening. In fact, in commercial offices, you already have 'light as a service', where energy suppliers will charge a monthly fee for the availability of light. The service provider installs and/or maintains the lighting, pays the electricity bills, looks after everything and promises you a specified amount of light, any time you need it, for which you pay a fixed fee.

Remember what I said before about how we pay for our smartphones? We don't pay for the phone; we don't pay for the texts, photos or emails we send; we don't pay for the data we use. We pay a fixed fee (based on typical use) and that's it – we just expect it to work. There are already innovative energy companies that provide energy as a service. They install solar and batteries and monitoring equipment at their cost, they charge you a flat monthly fee, and they guarantee you have electricity and heat when you need it.

The biggest disruption of all will come with autonomous

electric cars, buses and other modes of transport. (It will happen in advance of full autonomy but will become more prevalent as autonomy increases. Getting rid of drivers saves a high percentage of running costs.) There are already examples of cities where you can pay a monthly fee – based on typical use and time of day used – and ride any train, bus, tram or taxi within the city. Who will want or need a car, in an urban area at least? I love cars by the way, always have, and still lease one as I write, but it would be cheaper for me not to have one, and to get taxis or public transport everywhere. Once you add the convenience, the autonomous taxi will take you door to door, or door to station, or station to door, and with the even greater reduced costs, I will no longer be able to justify to myself why I pay the extra to have my own car, which sits idle 95% of the time.

The clean energy and clean transportation transitions will not be well under way by 2030 – rather, it will be 'game over' by then. Driving an ICE will be like going to a phone box with a handful of coins to make a call, or queuing at the bank to cash a cheque – and who still does either?

data networks will make autonomous driving ubiquitous and change the way we move almost overnight. Government policy will be the only hold-up here, not technology or cost.

As a service

Technology innovation is only half the story in any disruption; business-model innovation needs to happen as well. Two of the biggest disruptions we've seen in recent years have been how ride-hailing companies such as Lyft and Uber have disrupted taxis and transport, and Airbnb have disrupted hotels and room rental. It was the combination of technology

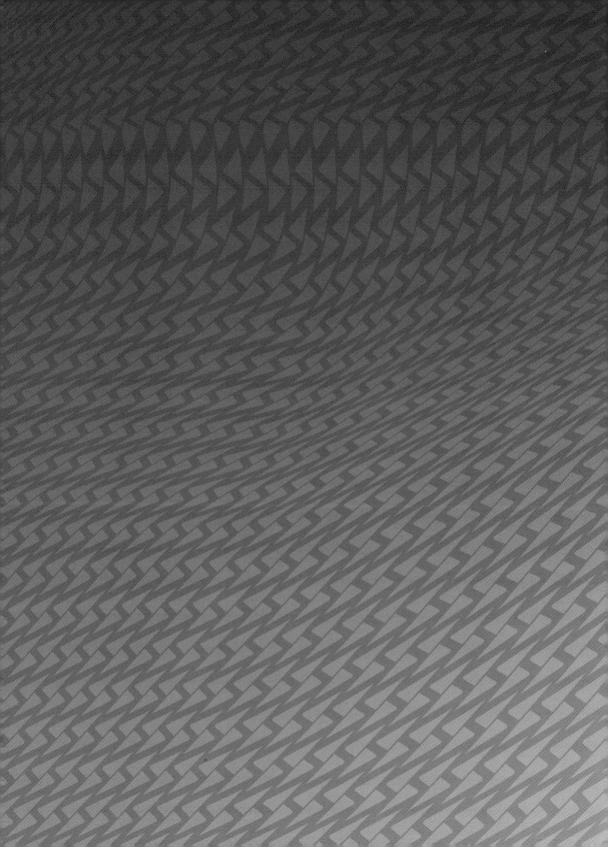

CLEAN ENERGY

SOLAR'S SHINING EXAMPLE
The persuasive case for photovoltaics

Nadia Smith

The next decade holds exciting prospects for solar to shine through, becoming one of the cheapest – if not *the* cheapest – sources of energy (surpassing fossil fuels). I remember being driven home from school, looking across the endless landscape and wondering if the rumours were really true – would we one day see fields utilising the sun's energy to power our homes and cars? We've come a long way; in the past two years, solar and wind have generated more power than nuclear in the summer months.

Projections from across the renewables industry show that, in most countries, solar photovoltaic (PV) will be the cheapest form of new electricity-generating capacity by or before 2030. Costs in the UK are expected to fall below £40 per mWh for large-scale projects, greatly surpassing previous forecasts.

This trend is due to a number of factors, although

Concentrated solar power (CSP) plant including Robert for scale!

an important key driver may be electric vehicles (EVs) themselves, or more accurately the increased co-location of technologies such as EVs, energy storage and solar. Within the next decade, affordable batteries are expected to become even more commonplace, helping to smooth out the daytime generation profile of solar power over 24 hours.

The UK's first 'subsidy-free' solar farm, Anesco's Clayhill Farm, was completed in 2017, co-locating 10 mW of solar PV with five energy storage units totalling 6 mW, marking a milestone in the solar landscape.

Electric vehicles are also on the verge of disrupting oil markets, with EVs expected to account for one in three cars on the road globally by 2040.* Future deployment of solar is likely to be offered on residential and commercial sites as a package alongside energy storage and EVs – echoing the movement towards more flexible and interconnected energy systems.

Exciting innovations in new solar technology also continue to spark the imagination of engineers turned artists around the globe. Trends in building integrated solar PV, such as solar roof tiles and clear glass panels, offer the possibility of turning skyscrapers into power stations or enhancing homes to become luxury low carbon living spaces. Large-scale applications also benefit from creative concepts that enhance efficiency, including bifacial modules – which can produce power from both sides of the solar panel – and the internet sensation the 'SmartFlower' – which uses a tracking system to follow the sunlight.

Innovators are also tackling the harder challenges that the industry expects to arise in the next decade: dealing with panel waste and end-of-life options. European project PV Cycle have developed a mechanical and thermal treatment process that can achieve a 96% recovery rate for silicon-based PV panels, with the remaining 4% utilised in waste-to-energy technology. They expect that by 2030, $450 million of recycled

Solar panels installed on roofs in South Australia

The SmartFlower: the world's first all-in-one photovoltaic system, pictured here in Le Havre, France (2017)

Trends in building integrated solar PV, such as solar roof tiles and clear glass panels, offer the possibility of turning skyscrapers into power stations

* *Electric Vehicles*, Bloomberg New Energy Finance, available at https://bnef.turtl.co/story/evo2018?teaser=true

Solar photovoltaics (PV) for power and solar thermal for water heating in Portugal

With high accessibility and ease of application, solar has become the favoured technology among communities and small-scale consumers around the globe

materials will be available, generating 60 million new panels from recycling (an additional 18 GW of capacity derived from recycled panels). These numbers signify a tiny percentage of the potential, as more panels reach their end-of-life after 2030.

And finally, with high accessibility and ease of application, solar has become the favoured technology among communities and small-scale consumers around the globe. The business models developed around solar, dubbed 'community energy enterprises', benefit from income from clean energy generation that is reinvested into the locality. These organisations also have a large focus on community ownership, often through

shares or another form of investment, and are engaging and empowering energy users in their consumption. The movement signifies a behavioural change, one where consumers become 'prosumers', with a vested interest in generating income by tackling the climate change agenda.

Not only do these groups exist in abundance in European countries, particularly Germany and Denmark (and the UK), but the decentralised nature of solar offers a great opportunity for off-grid applications in rural areas, especially in economically developing nations. In areas where the norm is reliance on a diesel generator or kerosene lamps, the low cost of solar is shining

through, offering safer, healthier and cleaner options to rural communities.

In more economically developed nations, community energy schemes are moving towards pairing solar with energy-efficiency measures – for example, the UK's Community Renewable Energy Wandsworth (CREW) cooperative have developed a model providing LED lighting as a service to community centres that would not otherwise be able to afford the upfront costs of purchasing the equipment. This helps decrease bills and tackles CO_2 emissions in the built environment, which currently make up 40% of the UK's carbon emissions.

Overall, the next decade looks bright for solar energy. With costs falling further, being supported by co-location of EVs and energy storage, we could see an end to the variable supply concerns that sceptics once identified as its downfall. But the innovations don't stop either: expect new markets for solar, with both creative technical developments and the growth of business models encouraging the behavioural change required for the transition to our future energy system.

One of thousands of solar PV installations retrofitted to a residential property in Bavaria

WIND OF CHANGE

Move fast, build things

Emma Pinchbeck

I don't know if you've noticed, but we've got wind (and I mean the meteorological rather than biological kind). The UK is a global leader in wind energy – a rare example, alongside queuing and sitting-down sports, of unquestionable British dominance that has grown up out of a mix of natural resources and our industrial heritage. We're the envy of the rest of the world in this vital low-carbon and cheap energy resource.

When I speak to people outside the energy sector, I'm surprised to find how many are still unaware of the success and scale of wind power in the UK. We've installed far more offshore wind capacity than any other country, almost 8 GW, and we also have 12 GW of onshore wind currently operating. That's enough to power over 14 million homes, providing more than 15% of the UK's annual electricity needs from wind alone. The most powerful turbines in the UK can generate enough electricity for an average home for a whole day with just one single rotation of their blades – each turbine is now a power station in its own right.

It would be a very contrarian energy analyst who didn't think that most of our power will come from renewable sources by 2030. Renewables are already generating more than 30% of our electricity a year. For one thing, we can see a massive pipeline of projects waiting for investment and construction, ready to go over the next decade. For another, we can see what the industry itself says about what's going to happen here in the UK: by 2030, offshore wind alone is looking to nearly triple in size,* providing over a third of our country's power from wind farms further out in the North Sea.

As offshore wind farms are some distance from the shore, it can be hard to get a sense of their scale. They're seriously massive bits of kit. The turbines are the height of the Shard skyscraper in London (approximately 300 m), and new blades on the turbines are the length of nine double-decker buses. In the UK, we've got seven of the world's top ten largest offshore wind projects. The

current global record holder, at 659 mW, is Walney Extension, off the coast of Cumbria in north-west England – although Hornsea Project One, being built off the east coast, near my mum's house in Lincolnshire, will be an even greater giant at 1,218 mW – making the walk we do to see it at Christmas even more fun.

All of this is not bad for an industry that only started building onshore wind farms in the UK in 1991, and offshore projects in 2000. The speed of innovation in wind is breath-taking, and a lot of it is being delivered by smart people in UK companies: we're making state-of-the-art turbine blades on the Isle of Wight; we're building *floating* wind farms off the coast of Scotland (tethered by cables to the seabed); we're using drones and AI to assess onshore turbines with pinpoint accuracy; and companies are making better components, like cables that can now carry

Robert chews the breeze beneath an array of wind turbines

It would be a very contrarian energy analyst who didn't think that most of our power will come from renewable sources by 2030

* 'Offshore wind energy revolution to provide a third of all UK electricity by 2030', available at www.gov.uk/government/news/offshore-wind-energy-revolution-to-provide-a-third-of-all-uk-electricity-by-2030

double the electricity capacity (66 kV instead of 33 kV).

Apart from the fact that engineering is cool, you should care about this – especially if you happen to drive an electric vehicle (EV) – because it has added up to staggering cost reductions on electricity bills. The cost of building and operating new offshore wind farms has fallen by about 50% in the space of two and a half years, making it cheaper than new fossil fuel or nuclear power stations. Onshore wind is even cheaper – in fact, it's

the cheapest source of new electricity bar none. The cost of onshore wind is so low that if the UK government were to invest in new onshore wind projects through its low-carbon auction system (something it hasn't done since 2015), it's been estimated that consumers would get about £1.6 billion back for their buck.*

As wind projects become this cheap, we are seeing businesses and investors fund some projects directly, without government intervention:

The cost of building and operating new offshore wind farms has fallen by about 50% in the space of two and a half years

Criticised by a small minority, but popular with the public, wind turbines are as majestic as they are essential

companies want to buy cheap, clean power to run their operations. But politicians will still want to invest our money in renewables for a couple of reasons. The first, as mentioned, is that we will make money back and reduce energy bills by having more cheap power on the grid. But the other major reason is to make sure that we get enough new renewable power quickly enough to meet our carbon targets and to meet rising electricity demand as we decarbonise our heat and transport by going electric. How much would it suck if we all moved over to EVs only to find out that we had to power them with dirty, expensive fuels?

The power sector is exciting, and what's particularly fascinating is how our new, clean, cheap power sources will interact with other changes in the energy system as we decarbonise. People are already switching over to new technology or asking for new services like energy tariffs that reward you for charging your car at a time that helps manage the electricity system cost-effectively – and this will be the norm for most consumers in the coming decades.

Renewables companies are especially interested in the explosion in energy storage technologies because they can use these to smooth out the differences in how they generate power day by day according to the Great British Weather. I'm currently obsessed by energy storage of all kinds, from big reservoirs that drop a lot of water through a turbine at the push of a button, to combining wind farms with their own batteries on site, to hydrogen, or even using EVs as a kind of portable battery. In this country, there are nearly 400 energy-storage projects up and running or in the pipeline. The next wave of battery-storage projects set to be built here will have enough capacity to fully charge 480,000 electric vehicles.

People used to think that because wind generates power according to weather conditions, it might present a risk to keeping the lights on. Now energy experts, including National Grid, are thinking that the best kinds of generation to meet consumer demand in this very different future energy system will be fast and flexible renewables and storage. It is great to be part of an industry that is powering our future. The Tech Giants like to talk about moving fast and breaking things. But for Cleantech Giants, it's all about moving fast and building things.

* *The Power of Onshore Wind*, BVG Associates, available at https://bvgassociates.com/the-power-of-onshore-wind

BOTTLING ELECTRICITY

Energy storage and batteries

Dr Euan McTurk

Ever since the dawn of the car, batteries have played a critical role in ensuring their success. The world road-speed record was set in the battery-powered *La Jamais Contente* in 1899, around the same time that New York had a fleet of electric taxis with battery-swapping stations. As road conditions improved and speed limits increased, lead-acid batteries became the Achilles' heel of electric vehicles (EVs), with their low energy-density meaning that range had to be heavily sacrificed for speed.

Ironically, it would be the lead-acid battery that enabled the electric starter motor for the internal-combustion engine (ICE) that relegated EVs to the sidelines for much of the 20th century. Batteries have been the limiting factor of EVs ever since, determining their range and charging times. However, the arrival of the lithium-ion battery in 1991, and its subsequent adoption and development by the consumer electronics industry, finally presented the automotive world with a battery that was fit for purpose.

The Tesla Roadster was the first mainstream electric vehicle to use lithium-ion batteries. In the days before lithium-ion cells had been mass produced in large formats, Tesla took the clever step of cramming thousands of small cylindrical laptop cells into the car, ensuring a steady and established stream of cell supply and providing a 240-mile range on a single charge, albeit for a high price. The more affordable Mitsubishi i-MiEV followed soon after, using 'prismatic' cells – big, chunky boxes that could be easily slotted into the vehicle – although it only offered a real-world range of around 60 miles per charge. However, as we have seen in the intervening years, battery capacities have rapidly grown, and prices have fallen. After nearly a century of largely stagnated progress on battery technology, how have these improvements been achieved so quickly?

The capacity of lithium-ion battery packs for EVs has been steadily improved by increasing the amount of 'active material' (the compounds that take part in the chemical reaction that produces electricity) in the cell, reducing the weight and thickness of non-active materials like the cell casing, module casing and internal separator between the positive and negative electrodes, and subtle tweaks in chemistry. These efforts have resulted in EVs being equipped with higher-capacity battery packs within the same physical space. By the end of 2018, Nissan had increased the battery capacity of the LEAF by 67% since its introduction in 2010; the Renault Zoe's battery pack capacity had been increased by 86% since 2012; and BMW had doubled the i3's battery pack capacity since its debut in 2013.

At the same time, battery costs have been decreasing as battery factories and raw material supplies are scaled up, and the amount of expensive material content, such as cobalt, has been reduced. For example, Nissan reduced the cost of the LEAF's battery from £205 per kWh to about £150 per kWh by the end of 2017. The battery industry is aiming to break the

Dr Helen Czerski gets to grips with batteries in an episode of Fully Charged

$100 per kWh barrier, at which point it estimates that EVs will not just be cheaper to run than ICE cars, but also cheaper to buy. Excitingly, Tesla and Audi claim to have secured battery costs that are tantalisingly close to this figure, meaning that such low battery prices could be widespread by the early 2020s.*

There's an exciting outlook for energy storage too. Cost matters more than density when it comes to home and grid energy storage, since the battery isn't confined to a finite

* Mark Kane, 'Audi: Claims EV battery costs of around €100/kWh ($112/kWh)', *InsideEVs*, 24 June 2017, available at https://insideevs.com/news/333577/audi-claims-ev-battery-costs-of-around-100-kwh-112-kwh

*Lithium-ion cells from left to right:
cylindrical cell, pouch cell
and prismatic cell*

space like a car. Bulky lead-acid batteries have been used for energy storage before, but they have a short lifespan and need to be replaced regularly. Lithium-ion cells have a much longer lifespan and require no maintenance, and as such have already proven to be successful in home and grid-scale energy storage projects. Additionally, there is an array of competing technologies on the horizon, including sodium-ion cells, flow cells and 'second life' EV batteries, which still retain at least 70% of their original capacity and therefore

have plenty of life left in them for energy storage. The breadth of different battery chemistries and the pace of their development should create an extra-competitive market and drive the price of energy storage systems down.

Even with today's technology and prices, batteries present a promising case for grid-scale energy storage. For example, the UK's National Grid would require the energy storage capacity of 10 pumped-hydropower stations the size of Dinorwig Power Station in Wales to fully cope with

a transition to renewable energy; that's roughly 100 GW hours of capacity.* At $100 or £78 per kWh, batteries could provide that amount of energy storage for £7.8 billion. Adjusting for inflation since Dinorwig's completion in 1984, the same amount of pumped hydro would cost £13.7 billion, and that's assuming that enough suitable sites could be identified.† Furthermore, as more and more households and businesses buy home energy storage because it makes economic sense for them, less public money will

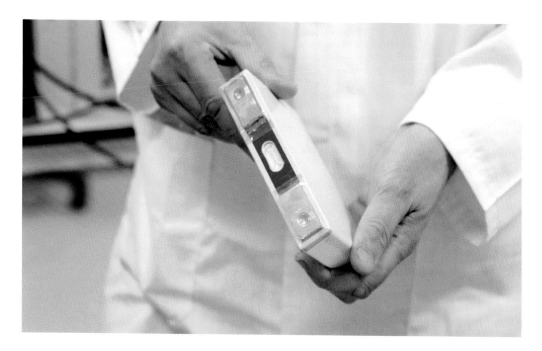

need to be spent on grid-scale battery storage, the speed of energy storage roll-out will increase, and individuals will directly experience the financial benefits of energy storage.

New battery chemistries will ensure a bright future for electric vehicles, too. Sodium-ion uses cheaper and more abundant materials than lithium-ion, while solid-state lithium and lithium-sulfur will provide lighter and more energy-dense batteries that are suited not only to long-range EVs but also to new applications such as electric flight. Advantages of different types of chemistry could be combined in the near future – for example, solid-state sodium batteries – resulting in everyday electric family cars that can travel upwards of 400 miles per charge, rapid charge in minutes and achieve cost parity with ICEs.

We're only at the start of the battery revolution, and yet EVs and home energy storage already make sense for the vast majority of people. Imagine what a no-brainer these options will be in five years' time.

* David JC MacKay, *Sustainable Energy – Without the Hot Air*, available at https:// withouthotair.com/c26/ page_191.shtml

† Elaine Williams, *Dinorwig: The Electric Mountain, A National Grid Publication*, 1991.

GREENING AND GROWING THE GRID

The grid will be reinforced by the energy transition, not weakened

Graeme Cooper & Pete Abson

Whether it's sitting behind the wheel for our daily commutes or travelling for holidays or days out with the family, cars are an essential part of modern life.

This has, however, also resulted in transport becoming the largest contributor to the UK's carbon emissions (27%), taking over from the energy industry (26%). Research shows that high levels of air pollution, to which traditional vehicles significantly contribute, is bad for the health of the UK's population, leading to 40,000 premature deaths and 6 million sick days each year, costing over £6 billion to the NHS and £22 billion to the UK economy.*

The UK government has recognised this in its case for change and commitment to accelerating the shift to low-carbon transport through its Industrial, Clean Growth and Clean Air strategies. However, there is a need now to speed up preparations for the uptake of electric vehicles (EVs) and to provide the right environment to support and encourage the consumer transition, if we are to develop a world-leading EV-charging network.

While interest in EVs continues to rise, the fear of running out of power while driving ('range anxiety') coupled with the lack of fast charging points across the UK is currently putting people off making the switch to electric. But this fear isn't unique to electric vehicles. Every day people driving up and down the country break down

because of empty tanks and misjudged fuel gauges. This is also a form of range anxiety. It is just one we have grown to live with, and stopping to refuel is now a normal part of daily life.

So, is range anxiety for electric vehicles more about perception than reality? And does the success in EVs' growth lie in accepting their limitations while working to improve infrastructure? A recent survey by comparethemarket.com identified that two thirds (66%) of motorists do not want to buy an EV or hybrid vehicle as their next car, citing range anxiety, the lack of charging points and upfront cost as the main reasons. A similar number (52%) expressed the worry that they would not be able to travel long distances as a serious deterrent to switching.[†]

Limited battery capacity and long charging times can deter drivers, but these issues don't have to be a barrier to accessing new technology.

National Grid have come up with a solution that will help open the world of EVs to longer distances and more cautious UK travellers. For two-car families, the first car covers an average of 37 miles a day, while the second travels a much shorter distance of 13 miles a day. The current 150–200-mile-range cars are adequate for this, meaning the answer to perception of range anxiety is not just bigger batteries but looking at how

Wind power: generating electricity till the cows come home

* 'Health damage from cars and vans cost £6 billion annually to NHS and society', University of Bath, 6 June 2018, available at www.bath. ac.uk/announcements/ health-damage-from-cars-and-vans-costs-6-billion-annually-to-nhs-and-society

† '16 million motorists say they can't afford to drive electric cars', 19 November 2018, available at www. comparethemarket.com/ media-centre/news/16-million-motorists-say-they-cant-afford-to-drive-electric-cars

products can fit in with our lifestyles.

Larger-capacity batteries add cost and weight, making them less efficient, whereas cheaper 200-mile-range cars could be the answer to mass-market uptake and avoid pricing families out of the market.

People are naturally cautious, and when investing in high-ticket items such as cars, we tend to plan for the biggest thing we do – such as the one long-distance drive each year, when a much larger battery would be needed than for the average daily commute. Consumers will only switch to an EV if they are confident that it will present minimal disruption to their daily lives.

National Grid are working to enable a smooth and efficient consumer transition to EVs and believe the answer to widening the use of EVs lies in investing in the infrastructure.

A network of car chargers, at appropriate speeds to suit the time spent at each location, will be vital to ensure we are ready for EV uptake. In addition to home, destination (gym, supermarket, etc.), local (equivalent to the local petrol station) and fleet charging, a network of ultra-rapid (120–350 kW) charge points is needed along the strategic road network, designed to make travelling and charging easier to fit in with long journeys and shorter rest stops.

When National Grid mapped the motorways in England and Wales, they found that 54 optimally placed locations would ensure that 99% of people using the road network would be within 50 miles in any direction of sufficient ultra-rapid chargers. This would avoid queues at peak times and enable drivers to charge their vehicles in around 5–10 minutes – not much more than the time it takes to fill up at the petrol station. This would help combat range anxiety.

National Grid already have the existing land and grid connections needed to install these charging points, and believe it is important to show that infrastructure needn't be a barrier to EV growth and that a structured and coordinated roll-out of rapid chargers is achievable.

This solution is about future-proofing so that, as the EV market grows, the infrastructure is in place to support it. It's about doing it once and doing it right, and taking a whole-system approach to both prepare for and invest in the future. With the right conditions, the UK has a real opportunity to become a global leader in electric vehicles.

> *Larger-capacity batteries add cost and weight, making them less efficient, whereas cheaper 200-mile-range cars could be the answer to mass-market uptake*

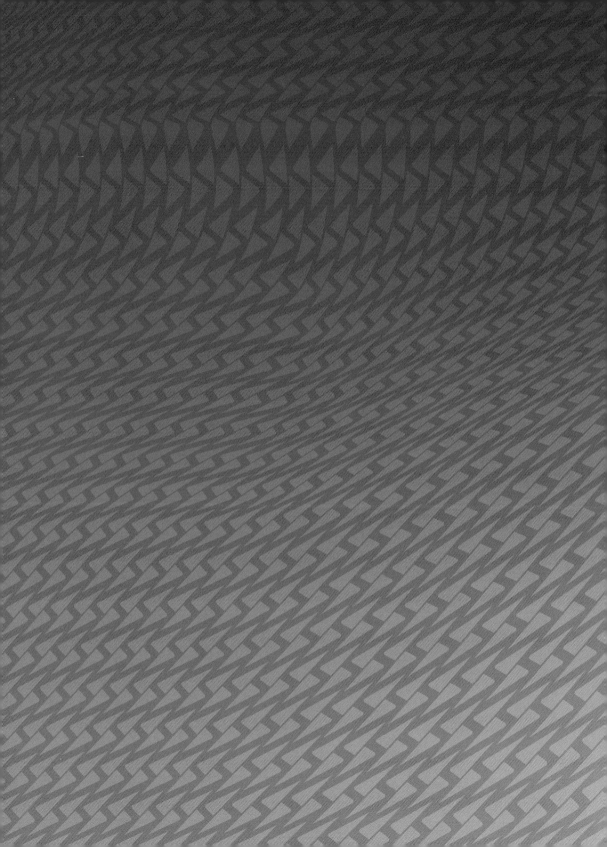

A FULLY CHARGED HOME

HOW CAN YOU MAKE YOUR HOUSE AN ECO-HOME?

Robert Llewellyn

For a moment, indulge me. Don't think about your house or apartment, or the climate, or the area you live in, or how much you pay for electricity. Just think about the impact new technologies for the 'built environment' featured on *Fully Charged*, and in this book, could have on your life. Let's look at the big picture for a moment.

If every building in the UK – not just houses, but schools, hospitals, offices, supermarkets, factories, warehouses, every available roof space – was fitted with now very cheap, very long-lasting solar photovoltaic (PV) panels, just how much electricity would that supply? It's in the terawatt-hour-per-year category – it would be huge.

It would also be very widely distributed, meaning National Grid wouldn't have to deal with one massive power station and the enormous expense of building high-voltage distribution from that site, as they are doing (at the time of writing) at Hinkley Point C in Somerset. Changes and additions to the grid would be required, but again not that substantial. Now, add batteries to all those buildings. Not necessarily huge capacity, but between 5 kWh and 10 kWh for a house, and obviously much bigger for hospitals, schools and industrial premises.

The total capacity of all those millions of batteries would be hundreds of gigawatt-hours. If they were all linked by a virtual National Grid, exchanging power as needed, the effect would be transformative, disruptive and very dramatic.

But would it provide all the electricity we need, day and night, year round? No, of course not. Would it produce a substantial percentage? Yes, without question, and as the UK is the windiest country in Europe, our offshore wind potential is more than enough to make up the shortfall. It is perfectly technically possible for this country to be 100% renewably powered, 100% energy independent, and reliant on no foreign fuels, with the concomitant benefits to our economy.

It would create an enormous number of well-paid jobs, increase the amount of money

Jonny and Robert, with Robert's solar PV, catching some rays while filming

staying in our economy and, of course, reduce our carbon output by more than anyone is currently suggesting.

Is it politically or economically possible? Of course it is, but it's not very likely. Will it happen? I genuinely think it will, but it will take a long time. Although there is an increasing need for all of us to release less CO_2 and mitigate the already profound impacts we are having on our planet, what will drive it is economics.

This is happening on a large scale now. Countries such as India and China have realised that solar and wind energy are cheaper. India has cancelled plans for dozens of new coal-burning power plants for purely economic reasons. Solar is cheaper. Why do we have

so many offshore wind farms around the UK? Because the power is cheaper.

So how can you start to do this in your home? That's a question an increasing number of people are asking, and I'm not going to lie, it's not easy. Yet. But here's one simple truth: it is perfectly possible today, and will be much easier in the next two to three years, to live in a house that:

» only uses electricity for light, heat and utilities like washing machines and fridges
» is able to produce up to 70% of its yearly electricity from solar and batteries.

All the technology and know-how exists and the cost of that technology is falling.

Of course, as many of us know, if your house has already been built it's not going to be so easy. Retrofitting an old house, finding a suitable part of the roof to mount solar PV panels, insulating the house to modern passive house standards, fitting ground source or air source heating, fitting a battery, installing LED lighting, removing the boiler, changing the water heating – all those things are costly and complex.

However, if you're building a new house, it's a very different matter. We are finally starting to see one or two new-build housing projects that are taking advantage of this in the UK, but they are pitifully rare. Instead, we are building tens of thousands of houses with

*A car being charged via a Rolec charger
from solar PV stored in a Tesla Powerwall 2*

lousy insulation, fossil-burning heating and no solar PV or electricity storage.

It is so mind-numbingly stupid. The building trade is 15 years behind the curve, very conservative, very boring and in dire need of major restructuring. It adds minuscule cost to the construction if new energy systems are integrated at the time of building, somewhere around 5% according to housebuilders I've spoken to.

As new systems of integration are developed, as doing this becomes the norm, that slightly higher cost of construction will disappear. The result is that this technology will reduce energy bills for the eventual occupants of the houses by around 80% a year. I'd say that is a very important selling point. If you look at the amount of money spent in the UK on new kitchens, a very popular home improvement, changing the focus to energy efficiency, electricity generation and storage is less of a big ask.

It's all about awareness and education. So yes, it's going to take ages and it's infuriating, but then so are a lot of things and you just have to breathe, try to explain and then move on. Thankfully, there are people all over the world who are refitting existing homes and

making a considerable impact. My family has done it in our house but it's taken years, cost thousands of pounds and it's possible I won't live long enough to reap all the financial benefits.

But then I am an early adopter, and as an American friend once told me, 'The early adopters are the guys who headed out West first, set the trail and made it possible for everyone following, but they got all the arrows in the face.' Yep, that's me.

Our house, built of wood in the early 1950s, has been majorly refitted in the 28 years we've lived here. It is incredibly well insulated, with triple-glazed windows and consequently low heating bills, but we still use liquid petroleum gas (LPG; we're off the gas grid due to our remote location) for central heating because I can't afford to fit air source heating yet.

The electricity we use comes from a combination of 5.5 kW of solar PV on a south-facing roof and a Tesla Powerwall battery system that can store 13.5 kWh. We heat water with excess solar and off-peak electricity in a Mixergy hot water tank. We use a Zappi charger for the car, which can, during the summer only, use excess solar energy to do so.

But here's the most important thing, and it took me a while to realise this: since we've had this system installed, we have used zero grid electricity between the hours of 16.30 and 22.00.

In the summer that's easy; there's plenty of solar power, and when the sun goes down we use the electricity stored in the Powerwall battery. In the winter the battery charges at night on off-peak electricity, and we use that power in the evening of the following day.

Why is this remotely important? Well, it's about reducing demand on the grid when electricity is expensive and dirty. Those two things always go together. Clean and cheap (wind and solar); dirty and expensive (coal and gas).

If you look at a graph of 24 hours of electricity consumption in the UK, from midnight to about 7 a.m. it's very low. It rises in early morning, when showers, kettles, washing machines and cookers are in use, then there's a bit of a dip, but in the afternoon, particularly in winter, there's a massive peak.

When there's a peak in demand, National Grid engineers call up all the excess generating capacity, mostly gas in the UK, and everything ramps up to cover the demand. What this means in practice

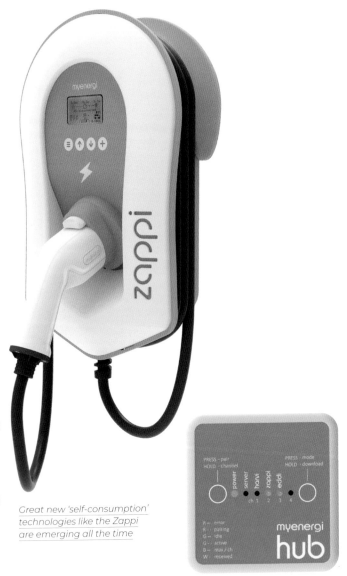

is there is a huge generating capacity that's on standby for most of the day. This is expensive: we have to pay to make sure that capacity is available at a moment's notice.

So, if nothing else, a battery in your house that charges either from night-time electricity or solar during the day would shave off some of that demand in the evening. One house makes no difference. A thousand houses might be possible to register. One hundred thousand houses would probably be measurable at the National Grid control room. A million homes, now you're talking – we could close down all coal-burning plants for good. We just wouldn't need them. Ten million homes in the UK – close down the gas plants. All of them, no more gas from Norway; sorry, guys, but I think you'll be okay. And the resulting cost benefits for us energy users would be noticeable.

In summer, my family's monthly electricity bills are always in single figures, and this is important because we also charge three electric cars. If we didn't have a car, our electricity bills would be zero from May to the end of September. There's irony. Get rid of your cars, Llewellyn!

So it's not impossible to make a big impact on your energy

Great new 'self-consumption' technologies like the Zappi are emerging all the time

consumption even if you live in an old building. If the house is being built now, it is incredibly simple and immensely cost-effective.

In the next three articles, we've gathered a handful of experts to tell you about some

of the things you might like to consider for your current home, like energy efficiency, solar and batteries, heat pumps and vehicle-to-grid. Plus, if you are looking to build an eco-home from scratch, we have something for you too.

STOP BURNING MONEY
Energy efficiency in the home

Tim Pollard

The subject of energy efficiency in buildings is very personal. It is defined by how and where we live, work and play. Inevitably, any article about energy efficiency must be general in describing both issues and solutions, so some may be relevant to you and some may not. However, there are things that we can all do, and some of the most effective strategies involve the single most complex component in any building – the people in them.

A good start is to understand how much energy we consume in our homes. Even here we discover that not everything is plain sailing since our energy companies describe our consumption in units called kilowatt hours (kWh). Many will struggle to understand what a kWh is and, of course, it is only useful to understand your consumption on a scale against others. Is your consumption 'good' or 'bad'?

Almost 80% of domestic energy consumption is a result of space heating (60%) and hot water production (20%). There are many things that will influence the energy consumed by space heating before we even start to consider the equipment. Notably, we need to understand how long during the day the building is occupied, how many people occupy the building and where the building is located (also, which way it faces).

The major priority should be to retain as much heat as possible by maximising the efficiency of the building's outer skin. This means insulating walls, roofs and floors as well as thinking about all the holes in the skin of the building to accommodate windows and doors. This should *always* be the first step in any strategy to save domestic energy. Cavity wall and loft insulation are both relatively inexpensive and they offer really good savings and a return on your investment. Double glazing and insulated doors are more expensive but still well worthwhile in terms of savings and returns. In homes without cavity walls, generally built before the 1930s, there are solutions available but they are much more expensive and intrusive.

Once the building is made as airtight as practically possible,

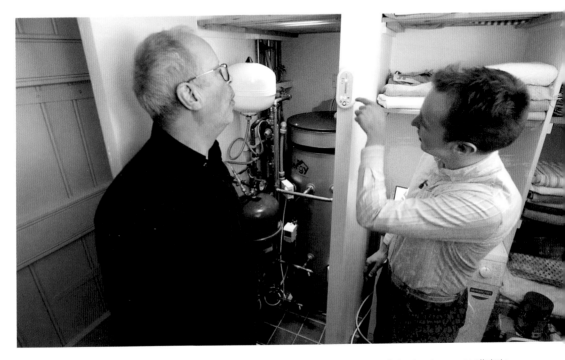

the next task is to look at the heating system. The system will generally consist of a heat engine (boiler), controls, pumps, pipes and heat emitters (usually radiators). Of course, the fuel used to power the heating has a large influence on the cost. Around 88% of us have gas-fired boilers, which are generally considered to be the most economic method of space heating. Since gas is approximately a quarter of the cost of electricity, the popularity of gas boilers has flourished. Modern boilers are more efficient than older models, but replacement on grounds of efficiency would not be sound economics.

However, there are other ways in which we can make our heating systems more efficient. Controlling your system can prove very effective economically and environmentally. We should aim to only heat our spaces when they are occupied and only heat them to a comfortable temperature. A 'full set' of heating controls comprises a room thermostat, a programmer and thermostatic radiator valves (TRVs). The room thermostat defines the target temperature, the programmer defines when the system comes on and goes off and the TRVs control the temperature for each radiator. Modern technology now allows remote access to your controls via the internet and a smartphone, and the most sophisticated systems have built-in smart features that 'learn' your preferences, detect presence and recognise weather conditions. A good set of controls is an excellent investment.

While boilers will mostly be replaced every 10–15 years, radiators are rarely subject to change. They are the forgotten part of the system but they can seriously damage its performance and efficiency. Additionally, limescale and sludge deposits can collect throughout the copper pipes. It is possible to manage this issue through the use of additives, filtration and treatment. Even the most efficient boiler will struggle if the distribution system is compromised. Maintenance of the system and the boiler should be regular and comprehensive and only carried out by a suitably qualified professional. The vast majority of boiler guarantees will be invalidated by a failure to have regular services and, at worst, safety may be threatened.

Don't forget the common-sense stuff too. Don't heat unoccupied rooms, turn the TRVs down to the minimum and don't obstruct radiators with furniture, curtains or anything else. Remember to change your control settings to reflect changes in occupancy times, weather conditions and individual preferences. The Energy Saving Trust have calculated that turning your room thermostat down by one degree could save £80 and 320 kg of carbon dioxide per year.*

I have concentrated on heating in the home but there is plenty of other stuff you can think about as well. Replacing older light bulbs with modern LED versions is quick and easy and can result in significant savings. Choosing the most efficient appliances is much easier these days with the introduction of energy labels and the now instantly recognisable rainbow scale. Don't forget to switch devices off rather than leave them on standby mode, as this can sometimes be quite punishing.

Lastly, remember that the cheapest energy is the energy that you never use.

> *Choosing the most efficient appliances is much easier these days with the introduction of energy labels and the now instantly recognisable rainbow scale*

* 'Thermostats and controls', Energy Saving Trust, available at www.energysavingtrust.org.uk/home-energy-efficiency/thermostats-and-controls

HEAT FROM THIN AIR

Cleaning up heat with renewable technology

Phil Hurley

I t's no secret that heating our homes can be both expensive and damaging to the environment. NIBE are the market leader for residential heating products in the Nordic countries, and we are committed to doing our bit in reducing carbon emissions and saving our customers money by staying at the cutting edge of clean energy technology with our ground- and air-source heat pumps. Heat pumps essentially work by transferring hot or cool air from one area where it is wasted to another where it is useful. Heat pumps have been used for decades within the refrigeration industry (obviously set only in the cooling mode) but are fully reversible so could be used to cool your house in summer and heat it up in winter. They create a perfect indoor climate by capturing natural energy from local surroundings.

The science

The science behind an air-source heat pump is relatively simple, involving two coils and a refrigerant substance. At one coil, the refrigerant is evaporated at low pressure when warm air is brought into the system via a fan. During the evaporation process, heat is absorbed into the refrigerant. Although extremely cold to us, this process occurs down to −20°C because refrigerants have such low temperatures that even this air is warm for them.

Then the refrigerant (and absorbed heat) is compressed, causing the temperature to rise even more. Once at the second coil, the substance is under such high pressure it condenses. When condensing, the heat absorbed earlier in the cycle is released into the central heating system to be circulated around the home or used to heat up water. An air-source heat pump absorbs heat from the outdoor air in winter to warm the home, and pumps heat out of the home in the summer to cool it.

Ground-source heat pumps work by drawing heat from the ground or ground water in a similar way to air-source heat pumps. Pipes buried in the ground or in nearby lakes or rivers extract heat from their surroundings. Underground temperatures are reasonably constant (10–12°C in the UK) so a ground-source heat pump, like an air-source one, can be used all year round.

Benefits

Heat pumps can lower your energy consumption significantly (80% in the UK for ground-source and 50% for air-source), providing cheaper energy bills and a more comfortable internal environment. When replacing high-carbon fossil fuel sources of heat, the carbon emission savings from heat pumps can be substantial, all helping to tackle climate change.

Heat pumps are a great way to turn your heating system green and can start saving you money immediately. They raise the energy efficiency of your house considerably and are easy to install during the construction of a property or if you are carrying out renovation works. An appropriately sized system can meet all your heating and hot water needs, and of course heat pumps don't require large deliveries of fuel or unsightly storage tanks.

Are heat pumps the future of energy?

Heat pumps are becoming more popular in the UK and across the world as the need to move to green heating

Heat pumps are big business in colder climes and are catching on elsewhere too

systems to tackle climate change is emphasised by governments and international bodies. Recently, the UK's Committee on Climate Change highlighted the importance of heat pumps, particularly for buildings that cannot currently be supplied by the gas grid and for newbuilds, which will consequently have a low-carbon, future-proof system in place.

In Sweden, France and Canada heat pumps are already a proven technology, and it is likely that similar growth will be seen in the UK and the rest of the world.

Norway, Sweden, France and Belgium, among others, have all made clear statements on the future of fossil fuels. The UK government has also committed to phasing out high-carbon fossil fuels in homes and businesses off the gas grid in the 2020s and recognises the strategic role that heat pumps could play in future heat decarbonisation.

The answer to the future of energy is right in front of us; we just need to utilise our natural resources in a smarter way. A range of solutions is needed, but heat pumps are definitely part of the mix!

CAR POWER FOR HOMES

Is vehicle-to-grid the next big residential disruption?

Mike Potter

There is a very important fact about renewable energy that eludes many people: it is rapidly becoming the cheapest form of electricity generation available. In the UK we have a lot of wind (energy, not political debate), which is low-cost but unpredictable, so having technology that allows electric vehicles (EVs) to use as much of this as possible is a true win-win. Cheaper electricity that is also greener.

The more we move towards a world in which we do not burn stuff to make energy, the closer we get to zero marginal cost electricity. Using EVs to coordinate energy use with erratic renewable generation helps to solve the big problems of this future world. You can quote massive numbers like

1.8 terawatt-hours of vehicle battery storage if we all had EVs, but first we have to make it work for normal users.

The idea of attaching EVs to the electricity grid and using them to help manage electricity flow is not new. In fact, Willett Kempton, the godfather of vehicle-to-grid (V2G), first wrote about this in 1997. Fast-forward to 2019 and we are on the brink of making this a reality for regular homeowners. Smart charging – where the charger is connected to the internet and can react to electricity prices – has received a massive boost in the UK as the grant system for home chargers now only supports chargers capable of smart charging (affectionately known as V1G).

Simply controlling when and how much electricity is drawn to charge a car not only avoids breaking the electricity network when we get millions of EVs on the road, but can also allow your car charger to 'buy' electricity when it's at its lowest cost. In conjunction with the University of Reading, technology company Crowd Charge calculated that for an average user this is a 30% savings. And here's the best bit: the cost of electricity is pretty much proportional to the CO_2 content of the generation, so you get 30% reduced CO_2 as a free accessory.

Much of this depends on the sophistication of electricity tariffs, and even though the UK is a leader in this area, I would expect to see New Zealand

The opportunity for electric vehicles to act as 'mobile batteries' alongside residential and commercial buildings is now being researched in earnest

The big news lately has been all about bi-directional or V2G charging

also benefiting soon with its fantastically dynamic electricity market, not to mention 80% of its generation coming from renewable energy.

But the big news lately has been all about bi-directional or V2G charging. With £30 million being spent in the UK and more on development and trialling projects across Europe and some small ones starting in Japan, hopefully the business case can be proven and this will in turn lead to mass-produced, suitably priced V2G charging equipment. In a world in which your car battery talks to your solar PV system, and the Zappi EV charger already talks to your home battery, talking to the electricity network cannot be far away – but what will it mean for the end user?

In Japan up to 6,000 homes have been using Nichicon bi-directional chargers for vehicle-to-home (V2H) for some years now, so the technology can be relied on. What will be important is how you use your car, how long it's plugged in for (the longer the better), how many miles you want to drive (the lower the better for V2G; the higher the better for V1G),

Floating solar array in a pond next to Engie headquarters in the Netherlands

Vehicle-to-building charger next to Engie headquarters in the Netherlands

what the rest of your house needs and most importantly how big your connection is to the electricity grid. All these factors will decide if you can get an annual savings on your electricity bill, ranging from £200 to £1,000, in a perfect case of correct car usage. These figures could increase over the next few years as up to 30 GW of wind power become available in the UK, so look forward to a bright future to save even more money on the cost of using your car.

Electric connections from battery system next to Engie headquarters in the Netherlands

ELECTRIC VEHICLES

ELECTRIC VEHICLES: WHY THEY'LL RULE THE ROADS BY 2030

Jonny Smith

If you have bought this book, or indeed if you're browsing its pages in someone's downstairs loo, presumably you already have an interest in electric vehicles (EVs). Maybe you already drive one.

The bizarre thing we – both the normal motorist and a lot of people rooted within the car industry – are still finding is that for some reason many folks still believe an electric car is a daft idea. They don't think it can do the job for the average motorist. I say average but no one really knows what that is.

These people are inherently sceptical. Their scepticism is holding them back from buying an electric car that 'still isn't affordable', even though there are plenty of sub-£30,000 EVs out there available on a personal contract plan with nominal monthly payments. Because who really buys a car outright any more?

The sceptics will have let the range figures scare them away from far cheaper running costs, less frequent servicing needs and the ability to use an app to warm the seats or pre-cool the cabin before setting off. In fact, the sceptics are still waiting for a £15,000 500-mile-range new battery car, by which point they've spent a fortune on fossil fuel, road tax and company car tax, and taken a hit on residual values.

I had a mate who was inherently sceptical of committing to buying a house. 'The prices will come down further and then I'll buy,' he said. . . for about seven years. The prices rose. And rose. And now he's living in a two-bed house when he could've bought a far larger three-bed if he'd not snoozed.

I'm saying all this but at the beginning of 2019 there were 5 million EVs on the road globally. Three years before that there were 1 million. China alone has more than 2 million passenger EVs, 400,000 electric buses, 5 million low-speed EVs and 150 million electric two-wheelers. China is the world's biggest economy, and China's tastes pretty much dictate what companies invent to then develop, build and sell.

Driven by its massive manufacturing colossus, China's cities have experienced a speeded-up version of the growth of cities elsewhere in the world. People flood to cities for jobs and the prospect of a better life. Cities get overcrowded. Factories and vehicles create excessive smog, while rising population demands more electricity for amenities and swelling industries. Governments realise this is unsustainable and vouch to clean up the air, so clean cars are welcomed and tail-pipe cars ousted. Further electricity demands drive the increase in renewable energy generation and encourage individuals to create their own power.

Boom. Well, not boom – actually more of a quiet buzz. Electric vehicles will be the dominant source of propulsion by 2030 because we are already witnessing a culmination of governmental laws focused on air quality, tax-based car engine emissions only rising, fuel prices only rising (watch oil companies as they continue to buy up alternative energy in order to future-proof themselves) and people increasingly being priced out of daily-use piston cars.

When I'm asked whether someone should buy an EV over a petrol/diesel car (which happens on a daily basis), I always tell people to think about what they use cars for. What's your average mileage? What are your typical driving distances day-to-day? Where do you live? Do you have a family? Are you a two-car family? Do you play the harp or own a St Bernard? (The last question referring to what boot space they might desire.)

Only after they've considered all these questions do I explain the running-cost benefits and

Jonny struggles to contain his enthusiasm for the engineering ingenuity behind the Rivian

cleaner-air benefits of an EV. Because in reality some of us care more about saving cash on a monthly basis than whether the air in our children's lungs is safe. Either way, buying an EV is a handy thing. My wife was a huge EV doubter, but within two weeks of living with an e-Golf she was habitually plugging in at home and enjoying the pre-heat function. It became no weirder than plugging your mobile in to top up its charge where possible.

The money she saved per month on fuel and road tax could be spent on clothes instead. Or a restaurant dinner. Or maybe some shonky old classic car for her husband to tinker with in the shed. More interesting still is that our children really enjoy the plugging-in ritual. They fight over who gets to do it. They also take an interest in seeing the range or the regenerative braking graphic displays on the dashboard.

Charging doesn't have to be a pain, especially as charge time will keep shrinking. In the same way that recycling in 1990s Britain was seen as pretty unusual and slightly inconvenient, it is now second nature. It is taught in schools, and the term 'upcycling' has led to young people embracing the 'make do and mend' mantra

of our grandparents – but with even more creativity.

My kids are under 10. They are yet to develop prejudices against new technology – in fact they will probably embrace electric vehicles, just like younger people accept and devour new phone/game/house tech and connectivity.

I have seen a lot of enthusiasm for Teslas grow, not from pure car enthusiasts but from tech enthusiasts who may not have been 'car people' prior to Elon Musk's transformative Model S and Model X. I see this as a good thing because it evolves the interest in cars and it fuels the dominance of EVs in our not-too-distant future.

Let's go back to the title of this piece – why EVs will rule the road by 2030. Because my children will probably pass their driving test in an EV of some form, and because when you speak to car designers and engineers they wax lyrical about how electric vehicles allow them more freedom.

Remember *The Jetsons*, the American cartoon depicting a family of the future, launched in 1962? They had flying cars, and George Jetson's didn't need parking at his office; he just pressed a button and it folded into a conveniently small

briefcase that he put on his desk. Well, obviously we're not there yet. Flying private cars would be a disastrous idea if you ask me, but last-mile flying delivery vehicles are being tested right now.

Electric cars have fewer parts to package, which means the shape of their cabin can be expanded and made more homely. Low-slung batteries and motors in skateboard chassis mean flat floors for more passenger room, better under-body aerodynamics and extra creativity for design aesthetics. The EV changes the proportions of the trad car as we know it without sacrificing occupant safety or comfort.

Besides new electric car production, I personally hope we will also see a resurgence of interest in hands-on vehicle customisation. The bare skateboard-type platforms may become available to buy from the likes of Bosch, General Electric, LG and Panasonic, which then allows the age-old coachbuilding business to be rekindled. Or young people could explore the notion of plug-and-play hot-rodding. Take a brand new skateboard EV chassis and bolt on the bodyshell of your auntie's old retro piston car instead of scrapping it. Or just 3D-print or mould a replica of a Ferrari body. A renaissance in home engineering would be no bad thing.

This upcycling enthusiasm has lots of scope. Naturally, with more EVs on the road there will be more used battery cells that can be made into home storage batteries – meaning home batteries will be cheaper and more commonplace. More recycled fibres and renewable materials will be hosted in EV production too.

With attitudes changing at an accelerated rate, the electric car is an ideal springboard or tastemaker for these ideas. Some people will want a conventional, conservative EV – which is absolutely fine. Toyota will probably make one to mirror their quietly world-conquering hybrid. But others will embrace the freedoms that EV design allows. Audi chose the e-tron to debut cameras instead of side mirrors in order to reduce drag. These will only get smaller. Look how Bollinger are utilising the lack of an intrusive engine to provide storage hatches through the whole of their electric sports utility pick-up truck. Rivian have a pick-up with a huge lower cargo deck that would have otherwise been taken up with 4WD drive shafts or

Electric cars have fewer parts to package, which means the shape of their cabin can be expanded and made more homely

exhausts. Both these vehicles can wade through deeper mud or water than existing off-road competitors. Will Land Rover step up with these plug-in threats to their USP? I have every faith.

The shorter-journey 'leccy delivery van is already being rolled out by Renault, Mercedes-Benz and Volkswagen. I personally think this will be the most important shift in electric transportation acceptance. Local plumbers in electric vans, the postal worker in a silent van. My increasingly habitual Amazon order will arrive in a semi-autonomous electric van. Imagine a tradesperson being given an electric van or pool car to use for their job. They may start out as a doubter but will soon see the benefits of EV life. What happens when they make a decision for their next personal vehicle?

Like it or not, autonomy lends itself perfectly to EVs. The hail-able pod-type taxi concepts emerging are all based on electric underpinnings. Tesla

brought OTA (over the air) software updates to car owners, and now others are following. The ability to customise or control your car via apps will work hand in hand with EVs. Sono Motors have developed a car whose body panels charge the car using solar energy – you can't tell me that won't catch on in some form.

So here we are, approaching 2020 knowing that Norway has pledged to ban petrol and diesel vehicles by 2025, as has India by 2030. Even the Scottish government said it will out-do the UK by phasing them out by 2032. Great Britain wants no new petrol or diesel cars on the road by 2040.

The evidence for EVs ruling the road by 2030 is stacking up nicely, and my hope is that convenient, faster charging opportunities continue to be explored by progressive small businesses, large corporations and collaborating car makers who realise that it is still a vital key on the chain.

My big question is this: what will happen to all the piston cars in 2030? Will mileage and fuel be rationed for classic cars? Will the world sprout more motor museums? Firing up a combustion engine could either become an occasional public spectacle or an illegal underground movement.

FULLY CHARGED

ELECTRIC VEHICLE MYTHS BUSTED!

'You have to regularly replace batteries'

Dr Euan McTurk

Electric vehicle batteries are lasting much longer than many critics originally anticipated. A prime example is Wizzy, the Nissan LEAF taxi from St Austell in Cornwall that clocked up over 170,000 miles on its original battery before being sold (not retired!). There are also many Teslas that have surpassed that mileage on their packs.

It's clear that EV battery packs last much longer than the lithium-ion batteries in phones and laptops, but why? This is in part down to far superior control electronics: the battery management system (BMS) that makes sure that the cells in the battery pack don't get too hot, overcharged or overdischarged. It's also due to their completely different operating environment: phones and laptops get very hot when in use, or even just sitting in your pocket, and are regularly left plugged into the mains while in use, which keeps the battery at its maximum voltage and causes the electrolyte to degrade over time. Conversely, EVs are unplugged while in use and spend most of their lives at ambient temperatures of 10–25°C, which is perfect for lithium-ion batteries in terms of maximising performance and minimising degradation. In fact, EV batteries last so long that there are widespread plans to use them in 'second life' applications, like grid storage, once they have eventually reached 70–80% of their original capacity – but this is taking much longer than their manufacturers had expected.

> *EV batteries last so long that there are widespread plans to use them in 'second life' applications*

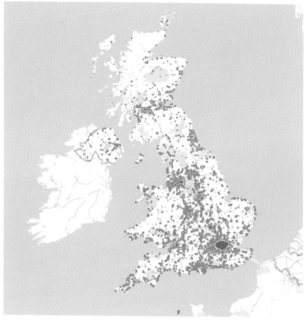

Above: ZapMap's map of public chargers in London
Right: ZapMap's map of public chargers in the UK

'The infrastructure isn't good enough'

Tom Callow

When people tell me that electric vehicle (EV) charging infrastructure isn't good enough, I usually respond with one word: 'legacy'. The infrastructure started to be rolled out seriously from 2010, when the UK government introduced the Plugged-in Places programme. By 2013, around 4,000 charge points had been installed, with local authorities owning many of them. While these have proved useful, it's fair to say that some of this legacy infrastructure hasn't been well maintained and is now less reliable. The good news is that private companies such as BP Chargemaster have been investing in infrastructure since 2011, and the majority of charging infrastructure being deployed today is privately funded, with no reliance on government support. Every week, more charging points are being installed, with a focus on rapid chargers that can charge vehicles as quickly as possible. Charging hubs – such as the UK's largest universal one, with eight rapid chargers, just off the M1 in Milton Keynes – are now appearing, and 150 kW chargers, which are three times faster than the current standard, started to appear in 2019. In short: we're getting more, better and faster charging infrastructure almost every day, and driving an EV is going to get even easier.

> *The majority of charging infrastructure being deployed today is privately funded, with no reliance on government support*

A renewable-powered Fastned charging station near Eindhoven, The Netherlands

'The grid can't cope with a surge in EVs'

Graeme Cooper

Electric vehicles will play a major role in delivering a low-carbon future. Through smart charging, vehicle-to-grid (V2G) technologies and consumers deciding to charge their vehicles at off-peak times, National Grid expect the growth in EVs to be manageable. In fact, it may be the case that at certain points EVs will actually be putting energy back into the system, so helping to balance supply and demand.

National Grid's 2018 Future Energy Scenarios analysis reveals that electric vehicles will be able to support the continued growth in renewables by storing excess generation and releasing it back onto the network when it is needed. The report, which provides an insight into how we could be using energy in the future and where it could come from, suggests electricity demand is expected to grow significantly by 2050, driven by increased electrification of transport and heating. There could be as many as 11 million EVs on our roads by 2030 and 36 million by 2040. Through smart charging technologies, consumers charging vehicles at off-peak times and through V2G technology, the increase in electricity peak demand could be as little as 8 GW in 2040. In turn, National Grid believe that EVs can support the roll-out of renewables by storing excess low-carbon generation and by providing electricity back to the system when needed.

There's a lot of work on infrastructure taking place across industry and the government at the moment, but in summary there will be a mix of charging: some people will charge at home or on the street, while others will charge at work, in car parks.

'You're just moving the pollution elsewhere'

Dr Nina Skorupska

Electric vehicles are only as clean as the grid that supplies them. It's our job to advocate to government a future where renewable energy makes up the majority of our energy mix. The UK has made excellent progress – summer 2018 saw a day where the majority of power came from solar, and there has been a host of coal-free days over the year. Power sector carbon emissions are now at historic, Victorian-era lows and are expected to fall to near zero by 2050.

Studies show that even when powered by the most carbon-intensive energy sources, electric cars produce significantly fewer greenhouse gases than diesel engines. A study by Belgium's Vrie Universiteit Brussel found that in Poland – which uses high volumes of coal – EVs produced a quarter fewer emissions than diesel vehicles over their lifetime.* Whitehall civil servants estimate that, in 2017, a battery electric car represented a 66% reduction in emissions compared to a petrol car.[†]

The benefits of electric vehicles are also multiplied when they are part of an interconnected system – a recent study from clean energy analysts BloombergNEF found that in the UK, EVs partnered with 'flexible' charging could reduce our reliance on gas-fired power plants to manage intermittency of renewables by 40% in 2030.[‡]

The uptake of EVs is part and parcel of wider energy system change and can play a key role in our transition to a clean, decentralised energy system.

* Arthur Nelson, 'Electric cars emit 50% less greenhouse gas than diesel, study finds', *Guardian*, 25 October 2017, available at www.theguardian.com/environment/2017/oct/25/electric-cars-emit-50-less-greenhouse-gas-than-diesel-study-finds

† *Transport Energy Model Report*, Department for Transport, available at https://assets.publishing.service.gov.uk/government/uploads/system/uploads/attachment_data/file/739462/transport-energy-model.pdf

‡ *Flexibility Solutions for High-Renewable Energy Systems*, Bloomberg New Energy Finance, available at https://data.bloomberglp.com/professional/sites/24/2018/11/UK-Flexibility-Solutions-for-High-Renewable-Energy-Systems-2018-BNEF-Eaton-Statkraft.pdf

Residential streets are about to undergo a subtle yet powerful transition

'No off-street parking, no EV'

Elinor Chalmers

Like one in seven people in the United Kingdom I live in an apartment, and my daily commute is around the national average of 20 miles per day. I am lucky enough to own my own home and a parking space, but having a home charging point installed turned out to be a logistical nightmare so I decided to go without. That did not stop me buying an electric car, however.

A pair of fast chargers and a rapid charger appeared in one of the multistorey car parks in Dundee, Scotland, where I parked several times per week. I had a lightbulb moment and realised I could charge while I was at work. Since then the local charging infrastructure has grown exponentially.

When you cannot plug in at home, you learn to pick up electrons whenever you are out and about. If I am passing a rapid charger I will stop for a 'splash and dash' for a boost to help me along my way. I tailor my shopping trips around local destination chargers. I will often ask permission to have a cheeky charge with my 24 kWh Nissan LEAF's 'granny cable' if I am visiting friends or relatives. On my way home from a longer trip I will often stop for a charge so I have enough for the next day.

> *Massive strides have been made in recent years to cut cobalt use in lithium-ion batteries, especially the ones used in EVs*

'Batteries are not ethical, recyclable or sustainable'

Dr Euan McTurk

This myth stems from cobalt mining in the Democratic Republic of the Congo (DRC). Cobalt is a key ingredient in the cathode (positive electrode) of many lithium-ion cells, especially the ones in smartphones, where cobalt comprises 39% of the materials used in the cathode (positive electrode) of the cell. However, massive strides have been made in recent years to cut cobalt use in lithium-ion batteries, especially the ones used in EVs: the latest EV batteries use cathodes that are made of only 4% cobalt, and Jeff Dahn's research group, who work closely with Tesla, recently found that cobalt can be completely removed from NCA cells, which are used in Tesla EVs. And let's not forget that there are many battery electric buses and trains (and some cars) that use a variant of lithium-ion battery called lithium iron phosphate, which contains no cobalt at all.

Where cobalt is still used in EV batteries, manufacturers are keen to ensure the use of suppliers with more ethical working practices. Mining companies are flocking to expand production in countries such as Canada and Australia, which have 250,000 and 1 million tonnes of cobalt reserves respectively.

Finally, lithium-ion cells are recyclable, and have been recycled for some time. The main drivers for doing so were the high market price of the cobalt and copper contained within them. Older recycling techniques were inefficient and struggled to recover lithium properly, but the latest techniques developed by companies such as Li-Cycle promise to recover all materials from the batteries – producing reusable battery-grade chemicals in the process – while vastly reducing recycling costs. The UK is investing millions of pounds into advanced EV battery recycling research through the Faraday Challenge, the UK government's programme to encourage research and development in battery technology.

'Over their life cycle, battery EVs are "dirtier" than fossil fuel cars'

Dr Euan McTurk

To show how wrong this myth is, let's compare apples with apples: the Volkswagen e-Golf, which does roughly 4 miles per kWh, and the Volkswagen Golf GTD BlueLine (diesel), which does 62.8 mpg and produces 189 g of CO_2 per mile.

Their chassis manufacture is a virtually identical process, so we can ignore the CO_2 emissions for both chassis since they cancel each other out.

For the battery manufacture and charging, it is important to look at the embodied energy – the energy required to produce the battery in the first place – since the CO_2 emissions will be dependent on the cleanliness of the grid. The e-Golf has a 36 kWh battery. A reasonable figure for the embodied energy of a lithium-ion battery is 556 kWh of embodied energy for every kWh of battery capacity, so that's 20,016 kWh of embodied energy for the e-Golf. This equates to:

» 16.4 tonnes of CO_2 if using only coal-fired electricity (highly unlikely)
» 9.8 tonnes of CO_2 if using only gas-fired electricity

» 0.6 tonnes of CO_2 if using only renewable electricity.

Tesla's Gigafactory in Nevada and BMW's Leipzig plant already run on renewables, and there are other battery factories built – or being built – in areas with very low carbon grids, such as Nissan in Sunderland and Northvolt in Sweden. Even in areas with carbon-intensive grids, installing on-site renewables at battery factories vastly reduces the embodied carbon of the batteries that the factories produce.

We know from Wizzy the LEAF taxi that EV batteries will happily last 170,000 miles, so let's assume that the total energy used to drive the e-Golf during its lifespan is 42,500 kWh (170,000 miles divided by 4 miles per kWh). As the UK's electricity grid rapidly moves away from coal, it's fair to take an average figure of 270 g of CO_2 per kWh of electricity. That equates to total driving emissions of 11.5 tonnes of CO_2. If charged using only renewable electricity, this figure is reduced to a mere

1.2 tonnes. Therefore, the e-Golf's total lifetime CO_2 emissions are a worst-case scenario of 27.9 tonnes and a best-case scenario of just 1.8 tonnes.

Conversely, the diesel Golf spews 32.1 tonnes of CO_2 out of its exhaust pipe over 170,000 miles, and creates at least another 2.7 tonnes during the refining of the diesel. This doesn't factor in the energy required to extract and ship the crude oil and ship the refined diesel, and also doesn't consider the air pollutants nitric oxide, nitrogen dioxide and sulphur oxides, and particulate matter emissions from the exhaust pipe.

Finally, the e-Golf's battery will still have at least 70% of its capacity left when it is retired from the car, which can then be used in grid storage to store renewable electricity and offset coal and gas plants at times of peak demand; diesel can only be burned once.

'Not enough choice and too expensive'

Mike Potter

By rights, electric vehicles should be cheaper to make than internal-combustion engine (ICE) vehicles. The considerably simpler construction of the EV drivetrain (10 moving parts compared to around 7,500 for ICE cars) means that it is only really battery prices that keep the sticker price of these cars high. According to BloombergNEF, as battery prices decrease, the cost of EVs will be equal to ICE cars by 2024.* But it's already way cheaper to own and operate an EV and here's why...

In fleet management circles people always talk about TCO, total cost of ownership. For the coming masses of EV drivers, here is some insight into the back-of-tofu-packet calculation that accompanies an EV purchase. (This works even better for used EVs – but there aren't that many around just yet.)

First we have fuel, the biggest running cost for most car owners. For an EV this is around £15–£20 for every £100 of petrol or diesel spent. So, counter to first impressions, you should buy an EV if you do more miles – the savings just keep coming.

The second-biggest cost is depreciation. EVs cost more to purchase, but after a bad start the second-hand values have improved. These could still be volatile as newer models are introduced, possibly at lower prices, over the next few years. At present EVs are marginally more expensive in depreciation terms compared to ICE vehicles. Expect this to shift dramatically over the next two to four years, as ICE vehicles are already getting harder to sell second-hand, and even if cheaper to purchase overall, the cost per month will be higher. You can always use lease costs as a guide to overall depreciation and interest, and with many vehicles you can be confident of the second-hand price.

The third-biggest cost is maintenance, and there is more good news here. Our experience is that EVs cost about one-third of a similar ICE car to maintain. All those ICE moving parts go wrong and need a lot of love and attention in the way of oil and checking. An EV's brake pads last three times longer due to regenerative braking. This is a major reason why used EVs make even more sense, as with ICE vehicles, maintenance costs escalate with mileage – this is not so much the case with EVs.

The last savings category is road tax, which is free on EVs, and there is a big tax saving on the horizon if you are a company-car driver (most people have given these up as... well, the tax is too high). In April 2020 the company car tax on pure EVs is due to drop – for something like a Kia e-Niro, from around £194 to £25 a month. Compare that to a petrol Niro, which bizarrely is £163 today (less than the battery car) but will keep going up over the next few years. On a Tesla Model 3 this would be under £40 a month.

So overall, EVs are already cheaper to run than ICE cars if you don't do really low mileage, and make sure you do some sums when you are shopping around – the sticker price is only part of the story.

* Jeremy Hodges, 'Electric cars may be cheaper than gas guzzlers in seven years', Bloomberg, 22 March 2018, available at www.bloomberg.com/news/articles/2018-03-22/electric-cars-may-be-cheaper-than-gas-guzzlers-in-seven-years

'EVs emit more pollutants than fossil fuel cars'

Dr Euan McTurk

This misconception is largely based on comparing large EVs to smaller petrol cars. When comparing apples with apples, or Golfs with Golfs, we find that the Golf GTD BlueLine weighs 1,320 kg compared to the e-Golf's 1,540 kg. This 17% weight difference may translate into slightly higher tyre wear, but is nowhere near the gulf in weight often used to claim that ICE cars somehow generate less pollution than EVs. As the energy density of batteries increases, this weight difference is set to fall. Electric vehicles rarely use mechanical brakes due to regenerative braking, which captures kinetic energy via the electric motor and uses it to charge the battery rather than using friction to slow the vehicle down. As a result, EVs very rarely need their brake pads or discs changed, and as such don't produce anywhere near as much brake dust as ICE vehicles. Furthermore, EVs do not emit any pollution from an exhaust pipe, while diesel vehicles pump out hazardous particulate matter, which causes serious respiratory and cardiovascular health issues at street level. These particles do not float off into the atmosphere but remain close to the ground, with increased concentrations at child and pet height. Coal use is rapidly being phased out in the UK, but when EVs are charged using coal power, the emissions from large, static coal power plants, located away from urban areas, can be more tightly controlled than the emissions from hundreds of thousands of small, mobile diesel engines. Most of the pollution 'from EVs' is actually from their tyres kicking up brake dust and particulate matter from ICE vehicles.

> *Electric vehicles rarely use mechanical brakes due to regenerative braking, which captures kinetic energy via the electric motor and uses it to charge the battery*

'Electric cars catch fire'

Dr Euan McTurk

It is actually incredibly hard to get an EV battery to burst into flames; I know because I've seen someone try (they had a good reason, honest). This usually entails abusing it by puncturing both the sturdy, secure outer battery pack cover and the individual casing of a cell, and then sticking a highly conductive metal object right through the electrodes in the heart of the cell. Even then, the cell may well not catch fire if it is not sufficiently charged. When a cell does catch fire, it typically gives plenty of warning beforehand, gradually getting hotter and hotter. This is picked up by the battery management system (BMS), which relays a

warning message to the driver, telling them to pull over and get out of the car, although to be fair any incident that is brutal enough to damage the car's battery – and all of the other car components that stand between it and the outside world – should be enough to prompt the driver to pull over anyway. A short while later (potentially several minutes), the cell will either just self-discharge – bricking itself in the process – and cool down again, vent gas or catch fire. The energy contained in individual cells in a battery pack is nowhere near as much as in a tank of petrol, so an EV fire is very slow-burning, with a number of

mini 'firecrackers' that give the occupants of the car plenty of time to get to safety. Compare this to the sudden, massive fireball that engulfs an ICE car and its occupants in seconds.

The press like to run stories on any EV fire that they can find because it's a relatively new technology and attracts a lot of attention, but don't forget that there are tens of thousands of ICE vehicle fires on the UK's roads every year, very few of which make the headlines. Above is a photo of an accident between a Tesla Model S (left) and a petrol car (right) in Luxembourg; which car would you rather be driving?

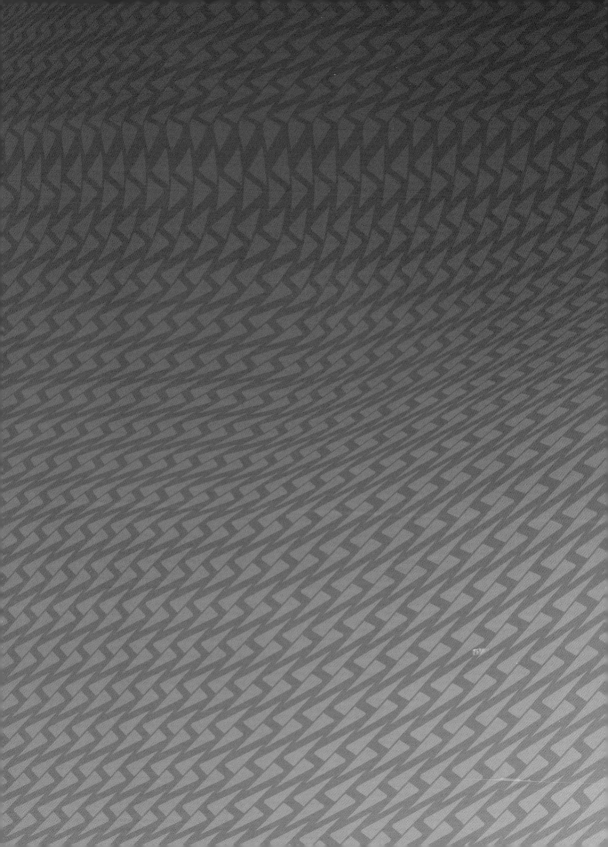

THE FUTURE IS ALREADY HERE, HERE & HERE

It's no longer California dreaming: electric cars and solar panels are commonplace on the West Coast

CALIFORNIA

Dr Nicholas Carter

When my wife and I moved back to California in 1997 after living in the United Kingdom, we test drove General Motors' cutting edge EV1. Actually, to be more precise, someone brought the sleek electric sports car to our house for a test drive. That someone turned out to be a salesman called Scott, who was incredibly enthusiastic about the EV1 and happened to mention that his personal goal was to become a NASA astronaut. So, not your average car salesman!

It turned out that we were in for quite the ride ourselves, as we embarked on the journey portrayed so accurately in Chris Paine's marvellous documentaries *Who Killed the Electric Car?* and its sequel *Revenge of the Electric Car*, much of which are set in California. Actually, if you look closely, you can see me in the former, holding my black-bordered EV1 picture at the Th!nk City protest.

Anyway, fast-forward 20 years to November 2017 and you find us taking delivery of our shiny new 300+-mile-range Tesla Model 3 in Fremont, California, where it was built. Our vehicle identification number is just over 1,000! As of writing (January 2019), we now have over 16,000 trouble-free, grin-inducing, smooth, clean miles on our Model 3 in addition to over 57,000 miles on our 2011 Nissan LEAF electric vehicle (EV).

Our home, in the San Francisco Bay Area, and our two electric cars are offset by a 3 kW solar electric system. As a sign of this burgeoning trend, we now have two near-neighbours who also have their own Model 3s, there are several Tesla Model Ss and assorted other EVs and plug-in hybrids on our street, and there's another all-EV household that actually has three EVs! Okay, so it's quite a long street but it really makes you feel that the future is already here. Quite a few of these homes also have roof-mounted photovoltaic (PV) systems, although it is the city of Palo Alto, south of us, that claims the highest PV watts-per-resident of any Californian city (second nationwide) at 4,748 watts.

California leads the United States in installed solar capacity by far, at almost 23 GW as of

June 2018. The next state on the leaderboard is North Carolina (4.5 GW). In fact, the next four states can only muster just over 13 GW between them. Also, cities in California hold three of the top four slots for solar jobs, as of 2017, with San Francisco far out in front at 24,474 jobs and Los Angeles second at 18,421. California also tops the league table for percentage of electricity generated by solar, at around 16%.

On the wind power front, despite being a pioneer, California is now number four, at 5.6 GW installed (almost 7,000 turbines), as Texas surged to 23.2 GW (as of the second quarter of 2018) and Oklahoma and Iowa moved into second and third place, with 7.4 GW and 7.3 GW respectively.

We are also seeing the growing trend of installing energy storage systems paired with residential solar arrays. As net metering of solar – which allows consumers who generate some or all of their own electricity to use it at any time, instead of when it is generated –

is reduced, storage is used to hold solar energy that would have been sent to the grid, so that it can be used to power a household during the evening, avoiding high peak electricity rates that still apply after sunset.

Looking forward, thousands more EVs will come out of Fremont, and perhaps from other new California-based EV manufacturers, plus electric motorcycles from local companies and thousands more electric scooters will be deployed in Californian cities.

The state has now committed to 100% carbon-free electricity by 2045, improving on its current status of around 56% and supplementing the renewable portfolio standard of 50% renewable by 2026. It has further committed to its economy being carbon-neutral by 2045. Given that California's economy is the fifth largest in the world, if treated as a country, this is a significant statement. It is also a major contribution to reducing our share of the world's carbon budget, which

the Intergovernmental Panel on Climate Change (IPCC) tells us, given business as usual, only has 10–12 more years to run until we face irreversible climate change.

It is interesting to think back to the time when we would wave at another new LEAF owner and excitedly point out an example of the latest model of electric vehicle; now EVs and plug-in hybrid electric vehicles (PHEVs) are commonplace, and California's car pool lanes – where EVs have single-occupancy access – are almost as busy as the regular lanes. Solar is also becoming a more common sight, although admittedly not up to the level of the one-in-five households seen in Australia or the one-in-eight of Hawaii*; still, that shows how much potential still exists.

Lastly, we are spoiled not only by our weather but also by the choice of food on offer, so that it is easy for Californians to eat a vegan, vegetarian or largely plant-based diet, enabling another huge contribution to greenhouse gas reduction.

* 'Net Metering Battles: Hawaii', energysage, available at https://news. energysage.com/net-metering-battles-hawaii

CHINA

Roger Atkins

China has become the pivotal market for electric vehicles, or, as they are known in China, New Energy Vehicles. It is the biggest market, the fastest-growing market, and the most integrated market in terms of government policy. There are several key reasons – both domestic and external – for this seismic shift away from fossil fuel mobility to the epic electron adventure now well under way.

Importantly to my mind, before we run through the key points, let's just compare and contrast the journey with the internal-combustion engine car outside of China.

Over a century of exponential growth, the motor vehicle has been at the core of how economies have developed and prospered – especially in Germany, Japan and the USA. It has had a massive impact on how towns and cities have been designed and operated; the 'car culture' has embedded itself into the fabric of society, significantly influencing where and how we work, what our leisure activities are, and often defining who we think we are by the car brand we choose to tie ourselves to. Gaining a licence to drive and acquiring a car have for decades been core rites of passage.

Meanwhile, over in China…

The 100-year journey the West took has been condensed into just a few recent decades in China as the country catapulted itself into a 21st-century consumer society. In terms of the auto industry, this famous quote from the early 1980s is particularly interesting:

Keep a cool head and maintain a low profile. Never take the lead – but aim to do something big.

Those words of wisdom are from Deng Xiaoping, the original architect of the astonishing journey China is now on – and the myriad consequences of his vision have been profound for both his country and almost every other person on the planet.

As I see it, Deng's 'something big' is the triggering of the mass adoption of electric vehicles, while 'New Energy', aka renewables, will be what they will all plug into. Batteries, among a few other energy storage devices, will be the coupling of the two movements. 'Keeping a cool head' and 'maintaining a low profile' have perhaps been the adoption of Western vehicle combustion engine platforms – typically previous iterations we've moved on from – and the establishment of joint ventures with most of the world's established players, while never entertaining a wholly owned foreign car venture. Notably Tesla's $2 billion Gigafactory in Shanghai is the first to break that rule.

The journey along the route set out by Deng Xiaoping has not been a silky-smooth ride, of course, as both rapid industrialisation and exponential fossil fuel car adoption triggered China's appalling air quality problem – as many as 29 million vehicles have been sold annually. The necessity both to acknowledge this issue and to deal with it sooner rather than later became a top agenda item for the Communist Party leadership.

Enter stage left: 'The Mother of Invention'.

Perhaps at this point it could also be said that 'the convergence of technologies' became 'The Father of Progress' in that several of

China's key internet players and their super-wealthy founders started to align their thinking around how things might be in the (near) future.

Baidu, Alibaba and Tencent were the tech titans that emerged as the Chinese wholeheartedly embraced the mobile telephony and internet marketplaces – they are collectively known as the BAT group. These operators are now key investors in and around the electric vehicle arena, doing so as we merge more and more with autonomous, connected and shared mobility platforms. There are other players now snapping at their heels as customers shift from being owners to users of products and services. Watch out for the TMD group going forward: news app Toutiao, group-buying service Meituan-Dianping and ride-hail firm DiDi Chuxing.

I'm often asked what's driving the shift towards electric vehicles. The answer I've been giving recently is threefold:

» Dieselgate: the revelation that a manufacturer had been in flagrant breach of trust with their customers.
» Tesla Model 3: the astonishing order book that was quickly realised against the odds.

» China's decree of an EV production quota from 2018 onwards.

I could write a tome on the first two but I'll save you the pain of reading any of that and cut to number three.

I shared my thoughts with Robert Llewellyn during our *Fully Charged* episode in July 2017 when I first spoke about the quota, which I see as the global trigger of change and the moment the electric vehicle 'renaissance' truly started.

In reality, it has a been a lesson learned from California, which is wholly apt given the

China is manufacturing and deploying solar at astonishing speed and in the strangest of places

electric vehicle journey that particular US state has been on since the late 1970s in its effort to ensure clean urban air for its citizens (big shout out to Mary D. Nichols of the California Air Resources Board for that).

Politically motivated, simplistic and unsubstantiated deadlines for the demise of diesel and petrol engines are popping up all over the world of late, but only in China have we seen the mandatory combined production of EVs and fossil fuel cars, with laws coming into place that effectively prohibit investment in new solely fossil fuel car factories. As we have been witnessing in many

European markets, EV demand is outstripping supply, and examining the reasons behind that will be very revealing in terms of policy versus profit.

China cannot and will not hope and wait for the electric vehicle market to kick off – the critical air quality issue is part of that rationale, as is the strategic objective of becoming an automotive leader rather than the slavish follower it has been to date. National pride and increasing capability are also key drivers, and we are certain to see increasing export efforts with both commercial and private EV brands very soon.

Given that much of the DNA of those products and brands is made up of both international and home-grown talent, we all have more to gain from increased collaboration while remaining cognisant of the competitive threat.

Moreover, the more we see China as a civilisation 5,000 years in the making, as extraordinary united states of diverse cultures and ethnic groups, as an opportunity to join forces with and embrace the global challenges of the 21st century, the more we will all thrive and prosper – and with the adoption of New Energy Vehicles at the forefront of it all.

COSTA RICA

Monica Araya

Costa Rica wants a life without exhaust pipes. Today an avalanche of cars drives around its cities, leaving behind the unmistakable smell of burned gasoline and diesel. So it is time to change that and the great news is that clean technology is on our side, national politics is on our side and many hearts and minds are going electric.

I never imagined that by 2017 I would be driving a car without an exhaust pipe in San José, the capital city. And little did I imagine that in 2018 our newly elected president, Carlos Alvarado Quesada, would embrace the notion that Costa Rica would get rid of fossil fuels and that he wanted this to be at the centre of his mandate. This does not happen every

day – and not in a world where we even see politicians denying the existence of climate change in the first place.

I have been an advocate of a fossil-fuel-free Costa Rica for several years and I don't remember a time like the one I see today. In the electric mobility ecosystem I have found a natural home because of the people who are excited by these disruptions, the many families who are proud to drive electric vehicles (EVs), the young people who love every piece of their electric bikes, and the many women who never cared about cars but are now EV activists. One vivid experience I had was when I presented an electric bus to amazed mothers and kids, who could not believe

wind. It's about 5–10 times cheaper to drive an EV than a fossil fuel car in Norway.

Today the latest-generation EVs can almost match fossil fuel cars when it comes to range, power and space. We still need a few more years before EVs can outperform fossil fuel cars in the same price range without any incentive advantages. But one thing is for sure: because of the increased numbers of EVs, Norway's car fleet has become cleaner, safer and less noise-polluting.

Famed for fossil fuels, Norway is divesting from oil and gas and investing in electrification instead

SCOTLAND

Elinor Chalmers

> *If you are catching a bus in Scotland, there is a good chance it may be hybrid or even fully electric*

Whether you are travelling around Scotland by road, rail or air, by 2030 your journey will very likely be electric.

The Scottish government has announced plans to phase out the sale of petrol and diesel cars by 2032 and has introduced an electric vehicle (EV) loan scheme to encourage drivers to take up greener motoring. This is ahead of the Westminster government plan to reach its goal of ending the sales of 'conventional' fuelled vehicles by 2040. By 2032, every town and village in Scotland should have public charging, via the ChargePlace Scotland network, with car sharing in place in most urban areas, reducing the need for car ownership. A key milestone that

Scotland is working towards is the commitment to introduce a low-emission zone in its cities by 2020.

Dundee, Scotland's fourth largest city, has just won the E-Visionary award from the World Electric Vehicle Association for being 'Europe's most visionary city for electric vehicles'. Dundee is leading the charge in Scotland by following a similar model to Norway. Three solar-powered charging hubs – one with battery storage – have opened across the city in the last year and serve the public and the over 100 electric taxis. One of the hubs has already been expanded due to high demand. EV owners are able to take advantage of free parking across the city. The local

council has the largest fleet of electric vehicles in the UK.

EV drivers in Scotland benefit from an active Electric Vehicle Association, EVA Scotland. Formed in 2011, they represent drivers at local and national levels and are part of the European Association for Electromobility (AVERE), where they support the drive for broader electromobility across Europe. In 2017 they developed ties with the world-leading Norwegian EV Association, with the parallels between the two countries allowing Scotland to take advantage of Norwegian knowledge and experience.

Initiatives by the Scottish government such as Plugged-in Households aim to grow public transport and car sharing, with the government giving access

to e-mobility car sharing based within Housing Association developments. Several community car-sharing projects have been running successfully across Scotland for years. Other projects are supporting the use of cargo e-bikes for city centre deliveries and e-bike sharing all round Scotland, supported by the Scottish government's funding for active travel and modal change.

If you are catching a bus in Scotland, there is a good chance it may be hybrid or even fully electric, especially if you are in Inverness, Edinburgh or Dundee, with more coming soon. Many of the buses are manufactured locally by Alexander Dennis in Falkirk and sold globally as far afield as Mexico and Hong Kong. In the area of autonomous transport, a self-driving bus service is planned from Fife to Edinburgh in 2020, using the Forth Road Bridge smart transport corridor, a service that could support 10,000 passenger journeys a year, helping to ease congestion on a busy route. The service doesn't add to the traffic in the

city, though, as it serves the tram–train interchange.

There are plans to electrify island-hopping around the Orkney by 2021. Flight times vary from the world's shortest commercial flight, 80 seconds from Papa Westray to Westray, up to 15 minutes from Kirkwall to North Ronaldsay. Cranfield Aerospace Solutions are working with Loganair to develop the first electric-powered commercial aircraft by modifying the regional airline's existing fleet.

Once you land in Orkney you will be amazed by the plethora of sources of renewable energy such as the impressive testbed of the European Marine Energy Centre, and the Surf 'n' Turf hydrogen fuel cell shore power system in Kirkwall that utilises surplus power on Eday to electrolyse hydrogen and offer a cleaner, quieter harbour. Also coming to Orkney is a world first in a hydrogen injection ferry. A living legend in the EV world in Orkney is Jonathan Porterfield of eco-cars.net fame (@ecocars1). He has almost single-handedly created the

Orkney EV community. Having recognised the advantages of EVs very early on, Jonathan realised the opportunity that electric power offered in an island group that has been a net exporter of energy for some time. For some of the islands, fuel is in short supply, while there can also be less road than range in most EVs. Canny islanders have been quick to realise the advantages of going electric in a place where pump prices are always high.

Scotland has produced many inspirational EV enthusiasts. Aberdeen-based Chris Ramsey and his wife Julie of Plug In Adventures (@pluginadventure) were the first-ever team to enter an electric vehicle in the 800-mile Mongol Rally. Dr Euan McTurk (@106Euan), a battery electrochemist and Dundee-resident EV driver since 2009, created the *Plug Life Television* YouTube series explaining how EV batteries work and their benefits, for all to understand. On social media, Elinor Chalmers (@She_sElectric) shares her everyday experiences of owning an electric car without having access to a charging point at her Dundee home.

There is no doubt that Scotland is leading the way to a greener future.

ENERGY & TRANSPORT FOR ENQUIRING MINDS

WHAT'S THE SAME AS A CHEESEBURGER?

How what we eat has its own carbon footprint

Maddie Moate

These days we're constantly faced with lifestyle choices that will affect our carbon footprint in a positive or negative way, but despite our best efforts to be carbon conscious and environmentally friendly, it's not always as simple as swapping this for that. We're bombarded with top tips on how to lower our emissions, but weighing up the options and calculating the actual cost of our footprint can be really complicated.

You should only purchase locally grown food… unless you're buying apples in winter, in which case definitely buy imported ones from New Zealand.

Stop using single-use plastic bags! But don't switch to a reusable cotton tote… unless you can use it more than 131 times.

Only boil the right amount of water you need in a kettle… unless you live in Iceland, in which case go nuts! Here your reliance on geothermal energy means your carbon footprint won't change no matter how much water you boil!

It's a minefield! To better understand my impact on the environment, I've calculated a handful of comparisons to give me some much-needed perspective, and for this, I've employed the help of a cheeseburger.
  A cheeseburger feels like a good place to start as the

shocking impact of beef (and dairy) on the environment has been well documented. Livestock production as a whole is responsible for around 20% of global greenhouse gas emissions, and when compared with dairy, poultry, pork and eggs, researchers found that beef is by far the worst culprit.

Producing beef requires around 28 times more land, 11 times more water and 5 times more greenhouse gas emissions than any of the other categories.*

Fully Charged presenter Maddie Moate on location with Robert

Robert reveals that he has 'accidentally' become a vegetarian, to someone's relief

* 'Land, irrigation water, greenhouse gas, and reactive nitrogen burdens of meat, eggs, and dairy production in the United States', PNAS, available at www.pnas.org/content/111/33/11996

The meat substitutes market is projected to be worth around £5 billion by 2023

One expert even suggested that cutting down on red meat would have more impact on our carbon emissions than abandoning cars!

These numbers are mind-blowing but it's still hard to appreciate the carbon impact of one person eating beef on the odd occasion. So, I've taken the CO_2e (carbon dioxide equivalent) of one cheeseburger and made a direct comparison with other approximate CO_2e measurements that my brain can actually visualise.

Assuming a single cheese-burger consists of a wheat bun, a beef patty, a slice of cheese, some token salad and a dollop of relish, it will have an approximate value of 2.5 kg CO_2e.*

But what's that the same as?

One cheeseburger (2.5 kg CO_2e) is about the same as:

» 1 kg of rice (10 average portions)
» 5 hot baths (heated by an efficient gas boiler)
» 48 cups of tea with milk (boiling just the right amount of water)
» using 5 new plastic carrier bags per week... for an entire year

10 cheeseburgers (25 kg CO_2e) is about the same as:

» 25 paperback books
» 50 pints of beer
» 313 bananas
» 1 passenger travelling 167 miles on a London bus

100 cheeseburgers (250 kg CO_2e) is about the same as:

» ironing 10,000 shirts
» a 42-inch LCD TV being used for 62 days continuously
» 1 person's clothing footprint for an entire year
» cycling from Land's End to

* All values are approximate and have been sourced from *How Bad Are Bananas? The Carbon Footprint of Everything* by Mike Berners-Lee, Profile, 2010.

John O'Groats (powered only by cheeseburger calories)
» a single flight from London to Glasgow

Whether these comparisons have helped or left your mind boggled, there's one thing for sure: the carbon cost of everyday stuff is complicated.

There are so many factors that come into play, and making the 'right' decision isn't always that simple. However, if we could all be that little more aware of our personal choices and their impact on the planet, I think we'd be off to a good start.

Robert tasted the Impossible Burger in Los Angeles and has never looked back

FULLY CHARGED FUN AND GAMES

Dr Kathryn Boast

Build your own battery

You will need:

» 1 lemon
» 2 pieces of bare, uncoated copper wire
» 1 galvanised (zinc-coated) nail
» Optional: voltmeter, LED

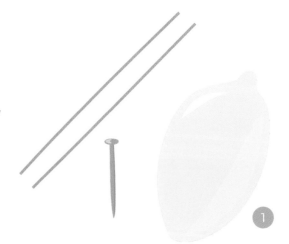

Start by rolling the lemon on the table, gently squashing it to break up some of the structure inside.

Twist one of the pieces of copper wire around a pen to turn it into a spiral. Screw this spiral into the lemon.

Stick the nail into the lemon too, close to, but not touching, the copper. This is your lemon 'cell'. Technically speaking, if you want a battery, you have to use more than one cell.

To investigate your cell, try touching the nail and copper

wire at the same time with one finger. Use an extra piece of wire to bridge the gap if you need to. Can you feel it tingle? Your finger is completing the circuit and allowing current to flow.

If you have a voltmeter, connect one terminal to the nail and one to the copper wire. What voltage do you get? This is the voltage of your battery!

You can hook up lots of lemons together in series to get more current from them.

Can you get enough current to power an LED?

What's going on?
In this battery, the zinc reacts with the lemon juice. These reactions drive electricity through the wire (or finger) that connects the zinc to the copper. The energy comes from the chemical change to the zinc.

Lemon juice contains citric acid and is the 'electrolyte' in our battery, connecting the two metals. The negative terminal, or 'anode', of our battery is provided by the nail. Galvanised nails are coated in zinc. In the acid, the zinc degrades and positive zinc ions (zinc atoms that are missing two electrons) float off the surface of the nail and into the lemon. The two electrons that are left behind then flow through the circuit connection (not through the lemon) and onto the copper. This is our electricity! On the surface of the copper, which is our battery's positive terminal, or 'cathode', the electrons join with hydrogen ions from the lemon juice electrolyte, producing hydrogen gas. This balances the flow of charge overall and allows our battery to keep producing electricity.

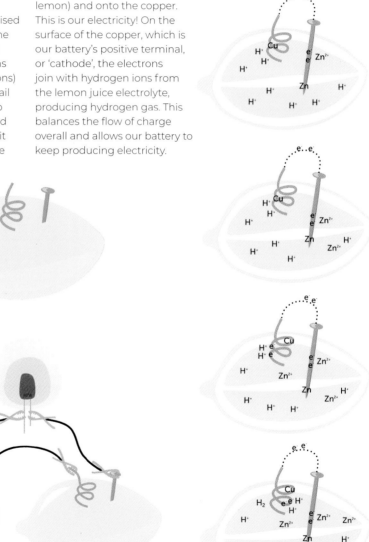

Make your own motor

You will need:

» 1 piece of bare, uncoated copper wire, roughly 1 mm diameter
» 1 AA battery
» 1 strong (e.g. neodymium), round magnet

Put the magnet on the flat end of the battery and stand it on the table. The magnet should ideally be just wider than the battery.

Bend the wire so that it balances on top of the battery and also gently touches the magnet. You can bend the wire into whatever shape you like to do this – try experimenting and see what works.

You should find that the wire will spin round the battery!

Don't leave it connected for more than about 30 seconds at once, or it may get too hot – we are short-circuiting the battery.

What's going on?

This kind of motor is known as a 'homopolar' motor. The electric current flows from the battery, through the wire, through the magnet and back into the battery.

When a current passes through a wire, a magnetic field is created. The 'right-hand thumb rule' will tell you which way the magnetic field goes – point your thumb in the direction of the current, and your fingers will curl round in the direction of the magnetic field.

The magnet also has a permanent magnetic field of its own, running from its north pole to its south pole.

The wire carrying the current will feel a force because the magnetic field that it creates will interact with the magnetic field of the permanent magnet. This interaction pushes the wire around the battery and magnet.

You can use 'Fleming's left-hand rule' to work out the direction of the force experienced by the wire. Line up your middle finger with the current in the wire, and your index finger with the magnetic field, which is at a right angle to the wire. Your thumb will then point in the direction the wire will move.

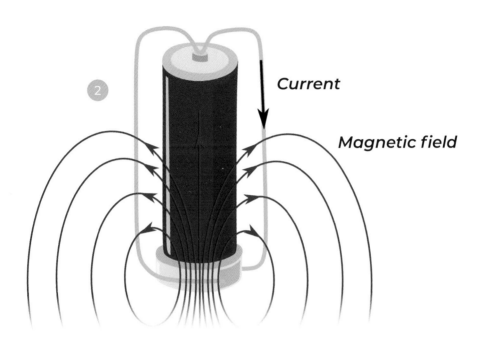

2

Current

Magnetic field

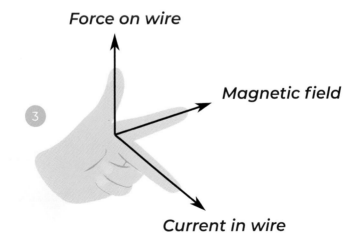

3

Force on wire

Magnetic field

Current in wire

One step further...

Michael Faraday was the first person to connect electricity, magnetism and movement, and he created the first-ever electric motor. You can reproduce his motor (with one important difference) using just a few simple bits of kit.

You will need:

» 1 9-volt battery
» salt
» 1 strong (e.g. neodymium) magnet
» 2 pieces of bare, uncoated copper wire, roughly 1 mm diameter
» wires to connect up the circuit
» 1 shallow dish

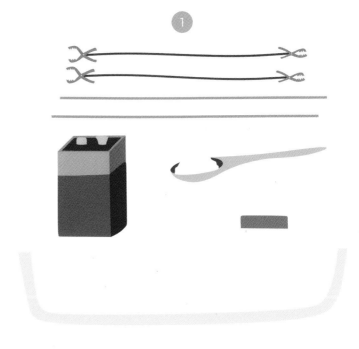

Start by making a saturated salt solution: dissolve as much salt as possible in water in the shallow dish. This salt solution will carry some of the current in our circuit. Faraday used mercury for this – so that's where our experiment differs!

Put the magnet in the middle of the dish.

Shape the two copper wires into rods with little hooks on one end. Hook them together, then use one rod to suspend the other over the centre of the dish so that the end is only just in the salt solution. You might want to stick the top rod onto something like a glass to hold it in place.

Using wires, connect one terminal of the battery to the salt solution. Connect the other terminal to the top rod.

The rod should rotate around the magnet.

If you can see little bubbles forming, the circuit is connected up correctly – you might need to give the rod a bit of a push to get it started if it's getting caught on something.

The same science is at work in a Faraday motor as in the homopolar motor we made on page 96 – the current flowing through a wire creates a magnetic field that interacts with the permanent magnetic field of the magnet, pushing the rod around.

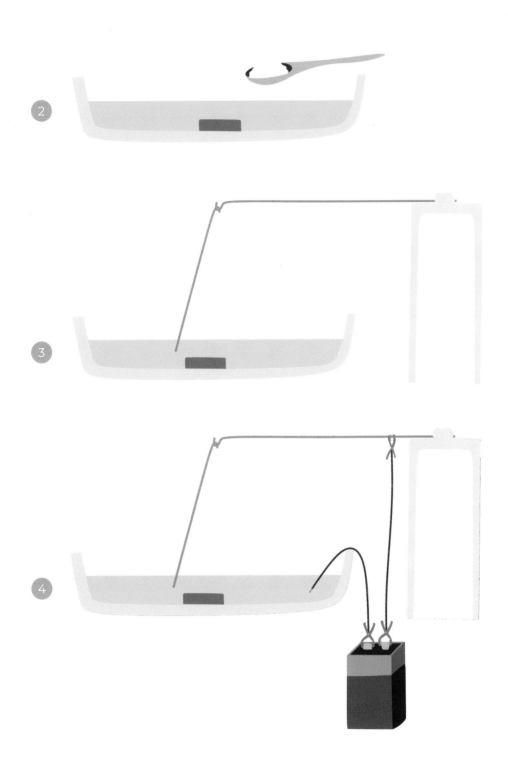

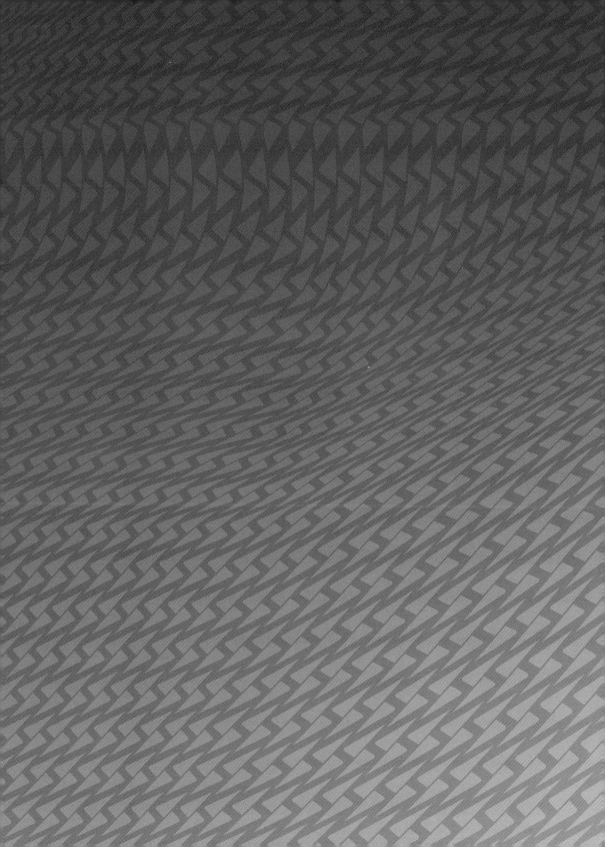

NEXT STEPS
FOR YOUR
JOURNEY

Robert and Jonny messing about on the canals of Amsterdam in an electric boat

LIVING A FULLY CHARGED LIFE

Angela Terry

Climate change and air pollution: these are topics that we are all hearing so much more about. It seems not a day goes by without a new report or survey telling us just how urgent the situation is. There is no question that radical changes to the way we live are needed. We need to harness and invest in the clean technologies that already exist. And there are plenty of other simple changes we all can make that are kinder to the planet and have the added benefit of saving money and making life easier.

'If you don't like something, change it' – it's a simple mantra that we can all live by if we have the knowledge and commitment to put it into action. But, let's face it, it's easier said than done. Changing habits is difficult. It's much easier when you can get excited about the results and the benefits that change brings. And this is where clean, renewable technologies come into their own. While the front-end investment for electric vehicles (EVs), electric bikes and renewable energy in the home may seem steep, people who change from being fossil-fuel-dependent to using clean technologies unanimously find that the investment is worth every penny.

Let's start with home – it's a good place to start as people's homes are one of the biggest sources of energy consumption. Installing solar panels is one of the most important things you can do to switch to clean electricity. It's a technology that is accessible to many homeowners, costing around £5,000, and the system will more than pay for itself over its lifetime. Remember to factor in rising electricity prices from your current energy supplier when doing the sums. In the UK, the cost of domestic electricity went up by 8% last year. If you take the plunge, using as much of the power generated by your solar panels in your own home rather than exporting the energy to the grid brings the best reward financially and environmentally.

If you have already put up solar, then what else can you do?

The biggest use of energy is neither electricity nor transport but heat. And the best way to stay warm, burn less fossil fuel and save money is through insulation. Wall and roof insulation will cut bills significantly and ensure you are heating your house and not outer space. It is recommended that roof insulation be at least 27 cm thick, which is nearly a foot, so it's worth getting out your ruler to check in the attic today. For cavity wall insulation, there is a 10-year guarantee and costs start from £100. Even solid walls can be insulated, either internally, which is the cheapest option, or externally.

If you're replacing your curtains or blinds, choose thermal ones and make sure they are drawn from dusk until dawn. Some simple DIY tasks can make a big difference – fix any windows or doors where the glazing is broken or hinges need adjusting. A simple installation of a draught excluder strip around doors and windows is really effective, as are chimney balloons if your fireplace is not used.

If you live in rented accommodation, you have a legal right to ask your landlord to make energy-efficiency improvements and they have to spend up to £3,500 if the property is substandard – an energy performance certificate rating of F or G. Not only does energy efficiency save you money on your energy bill, but

it provides you with a warmer and cosier home for years.

Upgrade your lighting too. LEDs save an amazing 85% of energy compared with halogen bulbs and the technology has really improved, so versions that dim are easy to get hold of. And when buying gadgets, choose the A++ energy efficiency criteria, and remember to switch things off when they are not in use.

Boiler upgrades can make a huge difference if yours is more than 10 years old. Heating accounts for about 55% of what you spend in a year on energy bills, so an efficient boiler is really important.

One of the simplest ways of supporting green energy is to switch suppliers to a green electricity supplier or tariff. Switching saves you £250 on average* and takes around 20 minutes, with no paperwork or engineer callouts. You cannot be disconnected for switching. It also means your home will

be powered by electricity from sustainable sources, such as wind and solar.

Not all green tariffs are the same. There are 'deep green' and 'light green' options here. Some suppliers source all their power direct from renewable energy projects, whereas others, the more 'light green' types, will just buy a certificate called a REGO (Renewable Energy Guarantee of Origin), which costs the supplier a tiny amount of money per unit. So as well as price and service, it is worth considering whether you are comparing green apples with oranges.

Review your water consumption too. Water use, its collection and treatment is an energy-intensive process, and simple measures such as installing a water butt for the garden and using low flow shower heads are worth considering to cut down on both water and energy bills.

Heating accounts for about 55% of what you spend in a year on energy bills, so an efficient boiler is really important

Renewable heat pumps work by using a series of heat exchangers to extract the heat from the ground, which is often at a stable temperature of 10–12°C. They do need a pipe network in your garden, which is either a borehole or slinky system depending on space and access, hence they have a fairly high price tag, but are a good technology nevertheless. Air source heat pumps work in a similar way but obviously the ground has far more stable temperatures than the air so the efficiencies are better for ground source heat pumps.

Next, let's look at what we sit down to eat. The production of red meat and dairy consumes a lot of energy and has a significant impact on the environment. Cattle and sheep belch out considerable quantities of methane, which is a harmful greenhouse gas. Simply switching from a latte to a flat white, and from beef to chicken or fish, can cut the carbon emissions of your meal fivefold, and can be cheaper and healthier too. Swapping rice for potatoes will cut your carbon footprint too, as rice production is a major contributor to methane and nitrous oxide emissions.[†]

But of course, all our food, indeed anything we purchase or consume, has a footprint on this planet, so buying 'mindfully' and minimising

An electric car in an electric train at the start of a 'fully charged' adventure

* 'What's stopping us switching energy provider?', MoneySupermarket, available at www.moneysupermarket.com/gas-and-electricity/switching-energy-provider

† Josh Gabbatiss, 'Rice farming up to twice as bad for climate change as previously thought, study reveals', *Independent*, 10 September 2018, available at www.independent.co.uk/environment/rice-farming-climate-change-global-warming-india-nitrous-oxide-methane-a8531401.html

waste is crucial. At the World Economic Forum in Davos in in January 2019, Prince William interviewed Sir David Attenborough and asked him what we could do to protect the planet. Attenborough replied that the focus should be on waste. Wasting energy, water, materials and resources costs us money and creates a pollution problem. Households in the UK alone waste £13 billion worth of food each year, so the trick is to buy only what you need and reuse leftovers.* But to complete the circle, add all the scraps to the food caddy collected by your council. The food waste is then broken down in a big digester without air and turned into biogas, which is added to the local gas network. This biomethane therefore helps to cook your dinner – a perfect circle of recycling.

Upcycling unwanted and unloved goods is another way to tackle waste – sell them on eBay or Gumtree, or send them to charity shops. Avoid black bin liners going to landfill where possible and consume

Switching from a latte to a flat white, and from beef to chicken or fish, can cut the carbon emissions of your meal fivefold

Robert on his electric-assisted bike in a low-emission city centre

less. TerraCycle has many recycling programmes that allow you to recycle a lot of your 'hard to recycle' plastics, which otherwise go to landfill, and the website gives details of local drop-off points.

And how do we get around our local area, dropping off our plastics, unloved goods and finding plastic-free options? It's difficult to justify doing all this good work while driving a fossil fuel car. The range of electric bikes on the market is impressive and there is a model to suit most needs – including cargo-style ones and ones that allow you to transport several children. Again, they require an investment, but they are still much cheaper than buying a second car and are fun to cycle too.

'Active travel' is the term used for getting around town without a car. It helps with fitness and weight control, cuts parking and fuel bills, and increases your feeling of wellbeing. If you haven't cycled for a while, the main thing is just to get on your bike.

One very well-documented way to cut carbon emissions is of course to cut down on the number of flights you take. This can seem like a big lifestyle decision but in fact there are so many advantages to staying closer to home – no airport queues, bugs or stress. Holidaying in the UK provides the opportunity to explore one of the most varied and captivating landscapes in the world. Love Home Swap, One Fine Stay, Seat 61, National Trust and Visit Great Britain are good websites to inspire your choice of UK holiday.

Finally, you can help plant trees. This is the only way to really absorb carbon out of the atmosphere and is a great gift to future generations – even if you never get to sit under them yourself.

* Rebecca Smithers, 'UK throwing away £13bn of food each year, latest figures show', *Guardian*, 10 January 2017, available at www.theguardian.com/environment/2017/jan/10/uk-throwing-away-13bn-of-food-each-year-latest-figures-show

With more electric cars on the market than ever before, the choice is yours!

HOW TO CHOOSE AN ELECTRIC VEHICLE

Jonathan Porterfield & Tim Shearer

So you want to buy an electric vehicle (EV). Good! Why? Is it because you want to do your part in helping to care for the planet? Is it purely for financial reasons? Is it a tax dodge? Whatever your reason, you probably have some questions: Where do I start? How far can I drive it? Which is the best EV for me? These are all excellent questions and the aim of this piece is to help guide you gently on your way.

Which EV is for me? Ask yourself, 'What is my main reason for needing an electric vehicle?' Do you drive 10 miles a day, 25 miles a day (the UK average) or 100 miles a day? Are you a commuter? Is it for the school run? Do you drive simply to pop down to the shops for a newspaper? These are all important factors in deciding what type of EV is right for you.

For instance, you don't need a 300-mile-range Tesla Model S if you only drive to pop down to the shop for your newspaper and a tin of baked beans, unless of course you're an A-list celebrity like Robert Llewellyn, in which case it's all about image. If you're not an A-list celebrity like Robert Llewellyn, a used four-seater Mitsubishi i-MiEV might be more suited to your needs. It has a usable range of around 60 miles and is 5% of the cost of a new Tesla.

The world's bestselling EV hatchback is the Nissan LEAF. Since 2011, over 400,000 of these vehicles have been produced. The LEAF has a range of battery sizes: 24 kWh, 30 kWh and 40 kWh models are available. These give a useable driving range of between 80 and 160 miles. The five-seat configuration makes this a good option for families. Other EVs in the five-seater hatchback category include the Kia Soul, Kia e-Niro, Hyundai IONIQ, Hyundai Kona, Renault Zoe and VW e-Golf.

Many parents insist on owning a massive SUV 4×4 to take their cherished little ones on the hazardous drive to school. This is reasonable as the school run can present many challenges. Herds of schoolchildren, prowling lollipop ladies and pothole-infested roads are the ever-present danger to the school run driver. Is there an EV to meet the challenge? The

Tesla Model X and Mitsubishi Outlander plug-in hybrid electric vehicle (PHEV) can comfortably take on the school run challenge. They're big, they're stylish and they can make the driver feel superior to other road users. Happily, both these vehicles can tow a horsebox (with a horse inside) so there's no need to fret – Felicity can still get to school during the week and enjoy her gymkhana at the weekend.

Charging

Now that you've chosen the EV that's best for you, where and when do you charge it? All EVs have dashboards equipped with information telling you the available driving range. The range will differ depending on your driving style and weather conditions. For instance, if you drive your EV like you're Lewis Hamilton, you're going to drain the battery and lower your driving range far more quickly than you would if you were to drive like Robert Llewellyn (or any other old bloke), who takes a more sedate approach. These scenarios are not unique to electric vehicles. If you drive a fossil fuel car aggressively, you'll burn through dino juice like it's going out of fashion. (By the way, it is!) You can charge your car at home from the same

socket that you use to charge your iPhone. It will take longer to charge than an iPhone and it's not as good at making phone calls, but while you're asleep your car isn't being used, so make that time count – charge your car. An interesting factoid is that 90% of the time 90% of cars are parked up, so remember the ABCs – Always Be Charging.

Early examples of electric vehicles from the first few years of this century (e.g. the G-Wiz) had limited range (20–25 miles) and speed limited to 30 mph. These may still be available to buy on internet auction websites but they are problematic and expensive to repair (often costing more than the value of the car), if parts can be sourced. The breakthrough for electric vehicles came in 2009 with the Mitsubishi i-MiEV and the Tesla Roadster, which both used lithium-ion batteries. These batteries meant that EVs could go further and charge more quickly.

Public charge points are becoming widely available, as are charging hubs where groups of EVs can congregate and graze on electrons while you shop. Often, the hubs are powered by solar panels. The cost of charging your EV at a public charging point varies but it always works

out cheaper per mile than fuelling an internal-combustion engine (ICE) vehicle with fossil fuel. Charging times vary from a 9-mile range being added per hour to a 150-mile range in 30 minutes. These charge times are falling rapidly, with Porsche planning a 62-mile range being added in six minutes for their range of electric vehicles.

A growing trend among EV drivers is to make their own fuel at home. Previously, for a driver of an ICE vehicle this would have involved many challenges. Firstly find a source of oil, secondly dig a big, deep hole in the ground, thirdly extract the oil, fourthly process and refine it, and then finally transport your highly flammable liquid product to some kind of holding station where you could pump it into your vehicle. By comparison, EVs are significantly easier to fuel at home. Place solar panels on your roof or install a wind turbine and hey presto! You're making fuel from the sun or the wind and are pouring it into your electric vehicle.

Financing

How to finance an EV? As with 19th- and 20th-century vehicles, there are various ways to own a car. There are the options to lease, contract hire, take out a personal contract plan or buy one outright with cold hard cash. Franchise car dealers are slowly realising the demand for electric vehicles and many smaller independent dealers now specialise in EVs.

So shop around. Request extended test drives and enjoy the experience. Social media is an excellent source of information from existing EV drivers. They're a friendly bunch and will be delighted to share with you their knowledge, experience and the smug feeling they get when driving past a filling station. Remember, there is no such thing as a stupid question!

Electric vehicles tend to have funky features not associated with ICE vehicles. For instance, watching your neighbour scraping ice and snow off his windscreen becomes infinitely more entertaining when you have told your car to heat up via the app on your smartphone. While still wearing your slippers and pyjamas, you chuckle at the age-old ritual of plumes of toxic smoke from cold engines ascending to the heavens as the neighbours cough and wheeze through the clouds of exhaust fumes.

In years to come, EVs will become more than a means of getting from A to B. They will be used as battery storage devices. What does this mean? At times of high or low electricity demand, utility suppliers will pay you, the car owner, to help support the local electricity grid by giving or receiving electrons to or from your electric vehicle.

Servicing

Forget oil, oil filters, belts, drivetrains, exhausts, clutches, ignition systems or the thousands of small moving parts in a fossil-fuel-driven engine. An EV service entails a check of the tyres and brakes and a pollen filter to make the car smell fresh. That's it! Oh, actually the screen wash will need to be topped up too.

The future

What can we look forward to? A growing choice of vehicles from all the major car manufacturers. Increased range with charge times dropping and the number of public charge stations increasing. Talk to any EV driver today and ask if they regret the choice they made in switching from ICE to EV. Driving an electric vehicle is nothing to fear. The future promises much but that doesn't mean you should wait. Technology will always improve, there will always be something new – take the plunge. Electric cars are not rubbish. The biggest change in the last 100 years of motoring has happened and EVs are here to stay.

In summary, an electric vehicle is cheaper to own, cheaper to maintain, environmentally more sound, sustainable and cool. The planet will never run out of electricity but it will run out of oil. Using oil to fuel vehicles will become the domain of classic car drivers who wish to look back whimsically at the past. Perhaps the time will come when the flammable liquids used to propel vehicles will only be available to purchase in shady backstreets from shifty characters who exchange jerrycans as they nervously look over their shoulders.

Henry Ford, the founder of the Detroit car manufacturer, once said: 'If I had asked them what they had wanted, they would have said a faster horse.' The point he was making was that sometimes people don't know what they want and it requires some 'outside the box' thinking to develop ideas that will change the world. The internal-combustion engine changed the world. Now electric vehicles are changing the world. So, what are you waiting for? Make the change – you won't regret it!

ELECTRIC CAR ROUND-UP

Robert Llewellyn & Jonny Smith

Aston Martin Rapide E

Jonny Smith

When CEO Andy Palmer moved from Nissan to Aston Martin, he took with him the vision of electrification. Aston's Rapide E – the pure EV version of their four-door Grand Tourer – will be on sale when you are reading this, but sales are capped to 155 to correspond with its top speed of 155 mph. Built in Wales using 90% British parts, the Rapide E was co-developed by Williams Advanced Engineering (they assemble the drivetrain in Coventry) and

promises over 602 bhp, 738 lb-ft of torque and 50–70 mph in 1.55 seconds. It weighs the same as the V12 equivalent (achieved with a carbon-fibre frame and the use of Kevlar) and the 65 kWh pack of 5,600 lithium-ion cells is where the engine, gearbox and fuel tank once were.

A dual rear motor set-up on the rear axle means it is 'proper wheel drive' and it boasts 800 volts and a 350 mph charge rate, allowing it to make use of forthcoming

350 kW DC chargers. Aston says their fast four-door has been developed for 'repeatable performance'. On the one hand this could look like a poor-value Tesla Model S, but on the other it is a far more exclusive proposition that allows buyer customisation and – most importantly – enables an Aston Martin to swan into central London free of Congestion Charge. This car paves the way for the forthcoming all-new Lagonda ultra-luxury EV limo in 2021.

Audi e-tron

Jonny Smith

Audi's choice to launch headlong into the EV world with an SUV is based purely around global trends. SUVs are the most in-demand car genre, and show no signs of slowing up. By making it largely familiar to look at, the idea is to encourage an effortless migration of owners from petrol and diesel vehicles into the new generation of Audis. Audi's four-ring badge holds huge kudos among owners, which I foresee will entice many new EV owners into the fold.

It is a shame the car isn't made of aluminium, and it's a shame that aerodynamically it doesn't match the I-PACE, but the Audi brings presence, seriously rapid charging and its familiar fantastic build quality to the table.

There will be another one or two cars to join the e-tron family in 2020, one of which will almost certainly be the Taycan-based 800 volt GT. The other will be a saloon called the Sportback, using the e-tron platform. Audi competing in Formula-E will only help to publicise their plug-in crusade.

BMW i3

Jonny Smith

It's hard to believe that this pillarless, rigid, carbon-fibre four-seater masterpiece first appeared in 2012. It still looks as fresh today as it did then – very few cars possess the style of an i3's minimal cabin. In fact, it was a generation before its time – most people weren't ready to accept it. From the bare carbon door shuts and skinny large-diameter alloys to the recycled-material dash and wool upholstery, the i3 was radical in looks but with obvious practicalities.

In 2019 the i3 REx range extender version was cancelled, with only the most subtle of facelifts the year before. Not that it needed one. Now there is a 120 Ah lithium-ion battery pack that delivers a range of 193 miles. The i3S is an even sportier version (the vanilla-flavour i3 is pretty engaging in all ways) that manages 177 miles. The ranges aren't headline grabbing, but it feels like you've been allowed to buy a concept car through BMW's back door.

There is no doubt in my mind that the i3 will go down in history as an EV classic.

BMW iX3

Jonny Smith

Whereas Audi and Mercedes-Benz were the first to wheel out SUVs, BMW – to our applause – focused on bespoke EV models. Now their established X model SUVs will emerge as EVs. The iX3 is first, with a twin electric motor 4WD set-up, and it goes on sale in 2020. Rumours of a 70 kWh battery and a WLTP (Worldwide Harmonised Light Vehicle Test Procedure) 250-mile range have been mentioned. Catering for the conservative EV buyer should only bolster BMW's EV portfolio.

A certain Mr Bobby Llew fell for this machine as soon as he clapped eyes on it

Bollinger

Jonny Smith

If the Rivian is 'dirty luxury' then the Bollinger is 'bare bones filth'. The Michigan-made utilitarian 4×4 looks like a Defender from a distance (and it's all aluminium like a Defender) but with over 600 hp, 668 lb-ft of torque and a 0–60 time of 4.5 seconds. The B1 (hardtop) and B2 (pick-up) models pre-date Rivian's products but they will emerge on sale at the same time. A certain Mr Bobby Llew fell for this machine as soon as he clapped eyes on it. The appeal is that the low-packaged drivetrain has created a front hatch to allow you to slide in and carry 5 m worth of, say, timber or carpet.

It has twin motors with a two-speed low-range gearbox and the 120 kWh battery provides a range of 200 miles. Four-door production starts in 2020, followed by the two-door and pick-up versions. Are we excited? Like a hog rolling around in sloppy effluent.

DS3 Crossback E-Tense

Jonny Smith

The Citroën luxury quirky sub-brand enters the pure EV arena with a premium baby five-door SUV that will share a Common Modular Platform (CMP) with the upcoming Vauxhall Corsa and Peugeot 208. Designed to rival Volvo's XC40 and Audi's A2 e-tron (due in 2021), the Crossback has a cabin that oozes tactile materials, quality infotainment and diamond shapes. Lots of diamond shapes.

The distinctly designed posh Citroën will have a 100 kW motor, 50 kWh battery, a WLTP range of around 185 miles, 80% charging in 30 minutes and 0–62 in 8.7 seconds. The bigger DS7 E-Tense will go on sale in 2020.

Honda Urban EV

Jonny Smith

2017's motor shows and car-based social media platforms singled out the Urban EV as one of the most inspiring and desirable cars. It didn't really matter that it was electric – people just loved it and wanted to see it in dealerships. It's won awards before it's even been launched, for goodness' sake. Not since the mk1 Insight has Honda produced any eco-car as interesting. At the time of writing, the Urban EV is due to go on sale in 2020 as both a three-door and five-door, complete with bijou lounge cabin and full-width dashboard screen. No performance or range figures have been released, but the author of this piece is close to slapping a deposit on one.

Honda messed up the hybrid game and lost to Toyota, so they need their mojo back and to be visible in the pure EV jungle. The Japanese car maker announced in late 2018 that from then on, each of their new model lines in Europe would feature electrified technology. Honda hope to have two-thirds of their new car sales in the region using electrified technology by 2025, five years earlier than their overall global goal.

Honda are an amazing company with a potted history of glorious experiments and characterful innovations. How many companies simultaneously develop robots that can run upstairs, championship-winning motorcycles and autonomous lawnmowers? Trust me when I say the world of vehicles needs more bold Hondas.

Hyundai IONIQ

Robert Llewellyn

One of my many problems, as lots of *Fully Charged* viewers will testify, is that as soon as I get into a new electric car it becomes my favourite, ever. It does not matter what I've been driving the day before or how much I've been raving about the such-and-such. This new one is the top shizzle.

That's how I felt when I got into the IONIQ. At the time it came out (2017) it felt immediately familiar, but better. It was, I think, the first time I had driven a 100% electric car that made me realise they'd become normal. I don't mean that as a put-down; I mean you could sit anyone with a driving licence behind the wheel and they'd start moving in 10 seconds. After a couple of minutes' driving they might say, 'Oh, is this car electric?'

That kind of normal. The good kind.

Hyundai Kona

Robert Llewellyn

The second day we had the Kona, I drove from my house in Gloucestershire (on the west side of the UK) to Nottingham (on the east side), a distance of 104 miles. It was a cold, dark evening in December. I was speaking on the hands-free phone for some of the journey, listening to an audiobook about the energy transition for the rest.

I got to Nottingham, delivered a talk, had some dinner and drove home. Driving is, let's be honest, for most of the time, essentially a very boring activity. You sit still and concentrate, for hours. That's what I did.

When I got home it suddenly struck me that I hadn't looked for a charger, I hadn't stopped to 'top up' on the motorway. I had driven just over 208 miles without a second thought. The car had 85 miles' range when I stopped.

The Kona is another massive, titanium-reinforced bolt holding down the lid of the scepticism-about-electric-cars coffin.

Hyundai Nexo

Robert Llewellyn

Contrary to many electric car obsessives, I love hydrogen fuel cell (HFC) electric cars. Not because I think they are better than battery electric, but because any type of technology that is a viable alternative to burning fossil fuels in an inefficient combustion engine is worthy of research and development.

The Nexo is without question a second-generation HFC car. With around 600 miles' range on one tank, it's a very viable alternative. I'm not going to go into the details of the need for a refuelling infrastructure (very expensive) or the downside of not being able to refuel at home, and the nagging question of where the hydrogen comes from, but this car is a delight to drive, easy to refuel should there be somewhere to do it, and would be, if it were available and people could afford it, an easy transition car. A petrolhead would be happy as Larry behind the wheel.

FULLY CHARGED

It's incredibly comfortable,
an amazing driver's car
with wonderful handling

Jaguar I-PACE

Robert Llewellyn

There's a phenomenon among gentlemen of a certain age. If they have been driving for two or three hours straight and they stop for a break, certain repressed grunting noises can be heard as they climb out of the vehicle.

Not so in the I-PACE. I drove for over four hours straight in the Netherlands in the I-PACE, got out at a fabulous Fastned charging station on the highway and didn't release even the most discreet of grunts. I was fine, I plugged the car in, it started charging at just under 100 kW and everything was dandy.

It's incredibly comfortable, an amazing driver's car with wonderful handling and very impressive acceleration, particularly if you're already moving. The jolt from, say, 50mph to 70mph is spectacular. Yes, it chews through electrons at a rather alarming rate, but I don't think the primary concern of anyone buying this car is going to be energy economy.

FULLY
CHARGED

> *The EQC has a beautiful cabin with huge screens and infotainment*

Mercedes-Benz EQC

Jonny Smith

Merc are not entirely new to the EV game, as some *Fully Charged* followers may know, or some may own the B-class Electric Drive. The B-class was a piston car adapted to EV and also hydrogen power (called the F-cell). The EQC is their first ground-up electric car in the EQ sub-brand, and like Audi they have launched with a conventional-looking SUV. It wears a distinctive front grille arrangement with fibre-optic illumination and some faux exhaust outlets at the rear – presumably so as not to scare off conservative buyers.

The EQC has a WLTP score of 280 miles and with an 80 kWh battery pack can rapid charge to 80% in 40 minutes. 4Matic means 4WD in Merc speak. Twin electric motors combine to make 408 hp and 0–62 in 5.2 seconds. The EQC has a beautiful cabin with huge screens and infotainment. By 2020 you'll also be able to buy a smaller version of this car, as Mercedes roll out more EQ products. Me? I'm waiting for a lovely 'leccy Merc estate car.

MG eZS

Jonny Smith

China is the world's largest market for electric cars, and MG are the latest manufacturer to enter, with a long-range electric vehicle with ranges in excess of 250 miles.

I have driven several of the new breed of Chinese MGs in the last few years, and they're much less rubbish than I thought. In fact, they made my face bend into a smile. The MG brand is highly revered in China – selling almost 200,000 cars in 2018 – whereas here I suspect we've fallen out of love with it since the dark days of the 1990s.

Nevertheless, the eZS is a small SUV (surprise, surprise) first seen at the Guangzhou Motor Show in China. No news on the details at this time, but the word on the grapevine mentions a front-mounted 148 bhp electric motor and battery good for a 268-mile range on the New European Driving Cycle (NEDC) test, the predecessor to the WLTP.

Mini Cooper S E

Jonny Smith

Electric-converted Minis and pure EV concepts have been bandied around for years now, but this could genuinely signal Mini getting their electric groove on. After all, the cars have got too big and less fun. Mini are rumoured to be bringing out an electric Cooper – or Cooper S E – based around sister company BMW's 42 kWh i3 undergarments, with 0–62 in under 7 seconds and a 200-mile WLTP range.

The Mini turned 60 in 2019, so it makes sense they would reveal the car in the anniversary year, to go on sale in 2020. Prototypes of the electric Mini were spied undergoing development testing in the Austrian Alps throughout early 2019.

Crucially a Mini has to handle like a Mini – massive grip and chuckability (that is a word, I think). This is where low-lying battery packs and motors will lower the centre of gravity and provide a highly entertaining hot-hatch driving recipe.

Nissan LEAF

Robert Llewellyn

In 25 years' time there will be a group of people who will meet in a park somewhere and open sandwich boxes and thermos flasks as they chat with each other. They will be members of the Classic Nissan LEAF Owners' Club. Their cars will be lined up under a solar canopy powering a line of charge points, the cars will all be immaculate and all will have driven several hundred thousand miles in their lifetimes.

The Nissan LEAF is a remarkable machine as it was essentially the first built-from-the-ground-up, mass-produced electric car that didn't cost £100,000. I've had one for nine years, and it's been 100% reliable and incredibly cheap to run. New tyres, a wiper blade and screen wash have been the only expenses other than electricity, and from May to September – thanks to solar panels and a Zappi charger – even that is free.

It's done 72,000 miles, and the average cost per mile? One penny. I love my LEAF.

Porsche Taycan

Jonny Smith

The Taycan – aka the lively young horse – is probably Porsche's most important model since the Cayenne was revealed to become their (bloody ugly but successful) cash cow. Thus far, £5.3 billion has been budgeted on the development programme. Gulp. It is a pillarless four-door sports coupé that has a similar silhouette to Porsche's current four-door Panamera, but thankfully with far better looks.

What do we know? Well, Porsche says it has to drive unmistakeably like a true Porsche. The underfloor batteries give it an even lower centre of gravity than their flat-six-engined 911 model, and the weight distribution will be 50:50.

Made from high-strength steel, aluminium and carbon-fibre, the Taycan will be around 4.8 m long and 1.99 m wide. In other words, 199 mm shorter but 53 mm wider than the current Panamera. The Tesla Model

S is both wider and longer. The interior is described as providing a typical 911-style driving position up front with adequate rear seating on two individual seats in the back.

There are no specifics on performance as yet, but expect versions between 400 hp and 600 hp, and all initially 4WD (in axle motors) with right-wheel drive coming later. We expect 0–62 in 3.5 seconds or less, with a range up to 311 miles using LG battery cells.

Interestingly, Porsche chose synchronous motors against the asynchronous motors favoured by Audi due to their ability to provide strong, sustained performance at high power – characteristics they say are key to the new car's development aims. In other words, they want repeatable high performance capabilities without the need to rest any part of the EV drivetrain.

The electric motors are similar in design to the unit employed on the petrol-

electric hybrid driveline used by the Le Mans-winning 919 Hybrid, with a solenoid coil featuring rectangular, rather than round, wiring. This has enabled Porsche to package the copper wires within the solenoid coil more tightly together to make the electric motors smaller than they would be using more conventional round wires.

Once it's revealed at the Frankfurt Motor Show in September 2019, it will be positioned price-wise between the Cayenne (£56,000) and the Panamera (£68,000), with plans to introduce battery-fed Boxsters and Caymans in future. The Taycan will be offered in two variants – the four-door saloon and the more rugged high-riding estate/crossover seen at Geneva 2018. Right-hand-drive Taycan cars will be delivered in Britain in early 2020. Tesla, check your mirrors, this could be the poster child of all performance cars in 2020.

This could be the poster child of all performance cars in 2020

Renault Twizy

Robert Llewellyn & Jonny Smith

Nothing we've driven in the last 10 years has made us laugh as much as the Twizy. Everyone who noticed the car as we pootled along in London smiled and waved. It's so ridiculous and yet it's actually very easy to drive and very sensible in a city. Parking a Twizy when you're used to trying to park a Tesla Model S is a non-event. You just stop and get out – it'll fit anywhere.

Here at *Fully Charged* we applaud Renault for bringing the bonkers Twizy to market all those years ago. It isn't exactly ideal for British weather, you can't go very far or fast in it, and it's not entirely comfortable, but it makes so much sense on small islands and congested warm cities, and our experiences driving it have been very positive.

For £6,500 it could be the perfect second car or urban weapon. The modern-day Messerschmitt bubble car. Will Renault follow it up with anything similar? It's getting a bit long in the tooth, so please say yes. Fun fact: in Japan these are badged as Nissans, thanks to the Renault–Nissan–Mitsubishi Alliance.

Renault Zoe

Robert Llewellyn

There are a lot of Zoes in London, and a ridiculous number of them in Paris – it's a really popular little car and the reasons are obvious. It's comfortable, easy to drive, goes far enough for 99% of people's journeys, is very energy-efficient and easy to look after.

I'm aware that in the UK there are some issues with charging infrastructure – it could do with a Combined Charging System (CCS) – but that's a very minor drawback when most charging is still done at home.

I haven't met anyone who's driven one who doesn't like it. It's not a performance car but it's nippy and agile, great fun to drive and the latest versions have very impressive range.

Rivian R1T

Jonny Smith

I hate the term 'game changer', so let's just say that Rivian's products have stolen the lion's share of headlines and promise to disrupt the traditional manufacturers. It can do 0–60 as quick as a McLaren or Lamborghini but is able to tow over 3,500 kg and carry more luggage than a van. Think Tesla meets Range Rover, so pure EV with off-roading capabilities never before seen in an electric vehicle combined with stunning luxury. When products like this sell for less than $80,000, I start to see the scale of electric car scope, and I see the acceptance of plug-ins even among the most staunch engine enthusiasts.

Rivian's CEO, R. J. Scaringe, is a visionary who set up the company at the age of 25. He has plucked engineers from established car makers and tech companies, and is about to sell one of the most disruptive EVs of 2020. The Rivian R1T double-cab truck has been so influential that Ford have reacted by promising a fully electric version of their most popular model – the F-150 pick-up. Which is ironic given that American-based Rivian tested their mules with F-150 bodies. This isn't the first time that the big, established companies are playing catch-up, but it's only a good thing to see. America, TRUCK YEAH.

It truly is the bubble car of the modern day, and is safer than countless cars that dwarf it

Smart fortwo EQ

Jonny Smith

I have a huge admiration for Daimler's Smart car brand. It truly is the bubble car of the modern day, and is safer than countless cars that dwarf it. The Smart EQ (formerly known as the ED) was the very first car I reviewed for *Fully Charged*, so it will always hold a special place in my heart. This generation of car was partnered with Renault. Smarts are not the cheapest cars, and by being two-seaters they rule out many buyers, but as an urban EV or for someone who lives in one of those ridiculously narrow Cornish fishing villages, it makes so much sense. My daughter is adamant this will be her first car.

Sono Motors Sion

Jonny Smith

A so-called disrupter new brand from Germany, Sono make what I believe to be an EV for the practical doer. In other words this is a small but capacious car that you can buy for €20,000. It comes equipped with towbar, external mains plug and an estate-shaped boot for plenty of cargo. I can see this becoming very popular for progressive retired people – the sort who own allotments or enjoy outdoor pursuits. The Sion can use its solar-panel-impregnated side panels to utilise free sunshine and charge up to 18 miles' worth of extra range while parked.

The battery has a range of 158 miles under the WLTP standard, with a capacity of 35 kWh. A single 120 kW motor powers the wheels. It uses efficient, monocrystalline silicon cells for the solar modules, and a total of 330 cells can generate a maximum power of 1,204 watts with 24% efficiency. The modules are attached to a substructure, similar to how a normal windscreen works, which means you can also swap the modules easily, should you experience some Parisienne-style parking.

There were 10,000 reservations on the Sono website at the start of 2019 so we can expect to see the first car in the world with a dashboard full of air-purifying moss imminently. But that's not even my favourite part of this car. The fact that Sono are openly sharing their assembly manual with customers online means that DIY-savvy owners are encouraged to repair or check their cars themselves.

Tesla Model S

Robert Llewellyn

I'm tempted to be really negative about the Tesla Model S just to stand out from the crowd of Tesla fan-people, but I can't do it.

This single car and the company behind it will be viewed in years to come as the car that turned the automotive industry on its head. I leased one for four years and I miss it every day; it's a fantastic car and everyone who's been in it is blown away by the performance, the big screen and the technology behind it.

It can drive itself, park itself, you can stand outside the car and park it using your phone. Okay, the last one may sound like a daft idea, but when you are in a British car park with an American car that was built for 10-lane highways and massive parking lots, you need all the help you can get. You reverse the car into the space from outside, so you don't have to try to squeeze out of the driver's door making unseemly grunting noises (see I-PACE for explanation).

The software controlling the car was updated countless times in the four years I had the Model S; the handling, fuel economy, car security and multiple systems were improved, and I didn't have to take it into a service centre to get those updates. They just happened, usually overnight.

The supercharger network set this car and the company apart from the start. By installing a hyper-reliable network of very fast charging stations all over the world where Teslas are available, they have kicked the sorry excuse of 'range anxiety' into the long grass – no, further, into a ditch that's about to be filled in by an electric excavator.

Yes, they are expensive, but there are clearly enough people with enough money to buy tens of thousands of them – they are now a very common sight on our roads. The very fast versions were another reason the other luxury car makers had to take notice. The Tesla Model S 100D Ludicrous makes a car costing twice as much, which needs expensive servicing every three months and drinks fuel like an open drain, look sluggish and old-fashioned.

Tesla Model X

Robert Llewellyn

As above but more embarrassing. Okay, maybe that's just me, but the rear doors... Try to get out of the back of a Tesla Model X discreetly. It's not going to happen. People stop and look at you. They may be thinking, 'Oh my Lord, they are so cool and modern and eco-aware.' Or they may be thinking, 'Pretentious tosser.'

That said, it's spectacularly comfortable, and it makes all combustion-engine SUVs look like a pile of sad museum exhibits, the ones in the corner covered in dust where the description label has fallen over because no one is interested.

I wouldn't want one but whenever I get the chance to drive one, I'm very, very happy to do so. And this car has a Llewellyn grunt factor of zero.

Tesla Model 3

Robert Llewellyn

The Model 3 in right-hand drive has been almost as eagerly awaited as *Toy Story* 4. It cannot come soon enough. It distills the DNA from its big brother Model S and makes it feel far more relevant, nimble and exciting for European drivers. Having driven a Model 3 in America, I was pleasantly surprised by its proportions, comfort and feel. Design-wise it's not exactly beautiful, but it's the packaging as a whole that really convinces me that this car will be seen a *lot* on British roads in 2020.

The £50,000 and £70,000 Performance Pack all-wheel-drive cars (0–62 in 3.7 seconds – oof) will emerge in right-hand drive first, followed by the rear-wheel-drive *c.*£30,000 version probably towards the end of 2020. One thing we do know is that Model 3s in Europe will come equipped with CCS charging ports.

Vauxhall eCorsa

Jonny Smith

An electric Vauxhall again – hurrah. When the second-generation Opel Volt/Ampera wasn't brought into the UK, nor a European badged version of the Bolt, it gave *Fully Charged* a sad emoji face. Maybe a slightly angry emoji face. Now Vauxhall have confirmed an electric version of their consistent bestselling supermini for 2020 production. With Vauxhall now part of the PSA Group alongside Citroën, Peugeot and Opel, the eCorsa will share major parts with the Peugeot 208 and Citroën C3. This means development costs can be shared. There are rumours of the next Peugeot 208 also being electric, including a GTi version.

The plans state that every Vauxhall model will be available with an electrified powertrain by 2024, with battery electric power for small cars, and hybrid and plug-in hybrid power for larger models.

No firm details have been revealed at the time of print, but expect the eCorsa to compete directly with the Zoe, LEAF and Kona upon launch. The Corsa consistently sells in the top five of Britain's cars, so an electric version will only widen the acceptance of EVs. Let's hope it's actually interesting.

Volkswagen e-Golf

Jonny Smith

The e-Golf was always planned as a stepping stone from the familiar electrified car into VW's ID range. The e-Golf does a convincing job of feeling utterly normal next to a combustion Golf, yet with a range of more than 130 real-world miles and excellent driving manners. It can never be as good as the ID, though, which supersedes it with the EV-specific MEB platform. The e-Golf won't be on sale in 2020 but the second-hand market will see plenty of them helping people get used to EV ownership from, cough, turbo diesel VWs. We almost wrote a whole paragraph without mentioning Volkswagen and Dieselgate.

Volkswagen ID

Robert Llewellyn

At the time of writing, this car still hasn't been delivered, but at the time of reading it will be hitting the streets. Not one or two, not in special showrooms where there's a nine-month waiting list. I predict this will be the first genuinely mass-market electric car that will be widely available, will blow people's minds and might just, if they are lucky, pull Volkswagen's reputation out of a toxic cloud of diesel fumes.

If you have ever driven a VW Golf, this is much better. It just feels solid, confident, agile; it has an amazing turning circle, it's incredibly easy to drive, it has an impressive range and it can charge very quickly.

I truly believe this car, along with the Honda Urban EV, will make electric cars normal and mainstream. We will start to see genuine mass adoption.

Volvo Polestar 2

Jonny Smith

The Polestar 1 plug-in hybrid electric vehicle (PHEV) was launched in 2019, so Polestar 2 will be Volvo's sub-brand's first pure EV. The Polestar 2 will be a mid-sized high-rise saloon priced between £30,000 and £50,000 to rival Tesla's Model 3. It will offer a range as high as 300+ miles, and as much as 400 bhp on tap. It is pretty, because Volvo's design language is smashing it out of the park right now. Their interiors are only a few notches down from Bentley's.

Volvo XC40 EV

Jonny Smith

China is the world's largest car market and Volvo are owned by Chinese giant Geely. In 2019 Volvo promised all their cars would be available as plug-in in some form, and the first full EV is their small SUV, the handsome XC40. This car was designed from the outset to house engines and electrification, so Volvo are making fairly safe steps into the electric market by using an existing car.

   The XC40 EV will launch after Polestar 2, Polestar being Volvo's performance arm.

ABOUT *FULLY CHARGED*

FUNDED BY PEOPLE POWER

Dan Caesar

One of the questions we are asked most often is: 'How does a YouTube channel make money?' and until now the answer has been: 'It doesn't.'

With funds from his own pocket for many years, Robert Llewellyn has managed to make *Fully Charged* episodes on a modest income of Google ads and the occasional sponsored episode. More recently, though, *Fully Charged* has primarily been possible thanks to the crowdfunding we receive via the Patreon platform. So it is no exaggeration to say that, without our Patreons, *Fully Charged* simply would not exist. In fact, just as importantly, Patreon has given *Fully Charged* the freedom to choose which stories we tell,

and the channel's popularity is in no small part due to the fact that its independence is very literally 'powered by people'. We are immensely grateful, and we will always need the support of like-minded individuals who, as we do, want to inspire the world to 'stop burning stuff'.

However, the global clean energy and electric vehicle revolution is gaining genuine traction, and the simple fact is that we are struggling to keep pace with the rate of change. Our small team can only shoot a fraction of the stories we would like to. This is why we need to start scaling up the business behind *Fully Charged*, but, critically, without losing what has made it such a huge success. To grow then, in addition to like-minded individuals, we also need the support of like-minded institutions, and that is why in 2018 we hosted the first-ever *Fully Charged* LIVE at the Silverstone Circuit in England. This exhibition proved incredibly popular, and as well as further establishing *Fully Charged* LIVE in the UK, we are working to replicate this success and grow our audience around the world.

To be clear, though, while contributions from individuals – whether it's for tickets or T-shirts or through

The Bristol and Bath Science Park, Bristol, home of Fully Charged

Jonny Smith, Helen Czerski, Robert Llewellyn, Maddie Moate and Dan Caesar at **Fully Charged** *LIVE 2019*

crowdfunding – remain core to the production of independent content, contributions from institutions will enable us to grow our team and grow our audience. What's more, profits from our commercial activities will go directly into growing our audience, and in turn this will attract contributions from new individuals. We believe that this virtuous circle will enable *Fully Charged* to be a sustainable, 'self-powering' – some might say 'self-charging' – business that will, we sincerely hope, inspire millions of people to switch to cleaner technologies.

GLOSSARY

Air pollution – the presence of chemicals or compounds in the air which are usually not present and which lower the quality of the air or cause detrimental changes to quality of life.

Alternating current (AC) – an electric current that reverses its direction at regularly recurring intervals.

Amp-hour (Ah) – a current of one amp for one hour. Often used to specify the current capacity of batteries. To work out the energy capacity, one needs to multiply by the voltage, e.g. a 12 V, 6 Ah battery contains 12 x 6 = 72 watt-hours of energy.

Arbitrage – buying a security in one market and simultaneously selling it in another market at a higher price, profiting from the temporary difference in prices.

Autonomous cars (or driverless cars) – cars capable of sensing the environment using a variety of sensors, and moving with little, or no, human input.

Battery – a container consisting of one or more cells in which chemical energy is converted into electricity and used as a source of power.

Battery electric vehicles (BEVs) – a type of electric vehicle that uses chemical energy stored in rechargeable battery packs. BEVs use electric motors and motor controllers for propulsion instead of internal-combustion engines.

Bioplastic – a biodegradable material that comes from renewable sources and can be used to reduce the problem of plastic waste that is suffocating the planet and polluting the environment.

Brake horsepower – horsepower measured at the output of an engine, i.e. before losses incurred in the gearbox and other parts of the drivetrain.

Carbon budget – an amount of carbon dioxide that a country, company or organisation has agreed is the largest it will produce in a particular period of time.

Carbon dioxide (CO_2) – a heavy, colourless gas that is formed by burning fuels, by the breakdown or burning of animal and plant matter, and by the act of breathing. It is absorbed from the air by plants in photosynthesis.

Carbon footprint – the measure of the environmental impact of a particular individual or organisation's lifestyle or operation, measured in units of carbon dioxide.

Carbon neutral – a term used to describe the action of organisations, businesses and individuals to remove as much carbon dioxide from the atmosphere as each puts into it.

Carbon offsetting – the action or process of compensating for carbon dioxide emissions arising from industrial or other human activity by participating in schemes designed to make equivalent reductions of carbon dioxide in the atmosphere.

Charge point – a mechanism for delivering energy to a battery, usually in the form of a high-voltage AC or DC supply.

Circular economy – an industrial system in which the potential use of goods and materials is optimised and their elements returned to the system at the end of their viable life cycles.

Clean air – air that is free of pollution, especially smoke or gases from vehicles, factories and power stations.

Clean energy (or renewable/ green energy) – specifically refers to energy produced from renewable or 'clean' resources without creating environmental debt.

Climate change – occurs when changes in the Earth's climate system result in new weather patterns that last for at least a few decades, and maybe for millions of years.

Coal – a combustible black or dark brown rock consisting chiefly of carbonised plant matter, found mainly in underground seams and used as fuel.

Cobalt – the chemical element of atomic number 27; a hard, silvery-white, magnetic metal obtained as a by-product from nickel and copper ores.

Community energy – low-carbon and renewable heat and power that is produced locally and distributed via a heat network or private wire arrangement.

Cost margin – the direct cost margin is calculated by taking the difference between the revenue generated by the sale of goods or services and the sum of all direct costs associated with the production of those goods, divided by the total revenue.

Decarbonisation – the reduction or removal of carbon dioxide from energy sources.

Decentralisation – the process by which the activities of an organisation, particularly

those regarding planning and decision-making, are distributed or delegated away from a central, authoritative location or group.

Diesel – the most common type of diesel fuel is a specific fractional distillate of petroleum fuel oil, but alternatives that are not derived from petroleum, such as biodiesel, biomass to liquid (BTL) or gas to liquid (GTL) diesel, are increasingly being developed and adopted.

Dieselgate – a term used to reference the actions of car manufacturing company Volkswagen, which had programmed vehicles to activate certain emission controls during testing. These allowed the vehicles to fall under the legal US diesel emissions standards during regulatory testing, while actually emitting up to 40 times the amount of nitrogen oxide the rest of the time.

Digital footprint – the information about a particular person that exists on the internet as a result of their online activity.

Digitisation (or digitalisation) – the conversion of text, pictures or sound into a digital form that can be processed by a computer.

Direct current (DC) – the unidirectional flow of electric charge. A battery is a good example of a DC power supply.

Disruptive technology – a technology that displaces an established technology and shakes up the industry, or a groundbreaking product that creates a completely new industry.

District heating – a system for distributing heat generated in a centralised location through a system of insulated pipes for residential and commercial heating requirements such as space heating and water heating.

Drivetrain – the mechanical parts that transmit power from the motor or engine in a vehicle to the wheels. The drivetrain does not include the engine or motor itself.

Economics – a social science concerned with the production, distribution and consumption of goods and services.

Ecotourism – a form of tourism involving visiting fragile, pristine and relatively undisturbed natural areas, intended as a low-impact and often small-scale alternative to standard commercial mass tourism.

Electric bicycle (or e-bike) – a bicycle with an integrated electric motor which can be used for propulsion.

Electric vehicle (EV) – a vehicle that is propelled by one electric motor or more, using electrical energy stored in batteries or another energy storage device. Generally used to refer to an electric car, it can also refer to an electric moped, motorbike, plane, scooter, train, tram and truck.

Energy as a service – energy service companies (ESCOs) develop, design, build and fund projects that save energy, reduce energy costs, and decrease operations and maintenance costs at their customers' facilities.

Energy efficiency – maximising energy usage; it is one method to reduce human greenhouse gas emissions.

Energy storage (or storage) – the capture of energy produced at one time for use at a later time. A device that stores energy is generally called an accumulator or battery.

Energy supply – the delivery of fuels or transformed fuels to the point of consumption.

Energy tariff – how an energy provider charges a customer for their gas and electricity use.

Energy transition – generally defined as a long-term structural change in energy systems.

Fabric first – an approach to building design which involves maximising the performance of the components and materials that make up the building fabric; this is before considering the use of any mechanical or electrical building systems.

Feed-in tariff (FIT) – a payment made to a producer of renewable electricity per kilowatt-hour generated. Designed to encourage the installation of small-scale electricity generation technologies, including domestic solar panels. In the UK the scheme was closed on 1 April 2019 and is expected to be replaced by the Smart Export Guarantee on 1 January 2020.

Flexitarian – a person who has a primarily vegetarian diet but occasionally eats meat or fish.

Fossil fuel – a natural fuel such as coal or gas, formed in the geological past from the remains of living organisms.

FUD – an acronym for fear, uncertainty and doubt.

Fuel cell – an electrochemical cell that converts the chemical energy from a fuel into electricity through an electrochemical reaction of hydrogen fuel with oxygen or another oxidising agent.

Fuel cell electric vehicle (FCEV) – a type of electric vehicle which uses a fuel cell instead of a battery, or in combination with a battery or supercapacitor, to power its on-board electric motor.

Giga (G) – times 1 billion, e.g. 1 gigawatt (GW) = 1,000,000,000 watts.

Greenhouse gas – a gas that contributes to the greenhouse effect by absorbing infrared radiation. Carbon dioxide and chlorofluorocarbons are examples of greenhouse gases.

Grid, the – an interconnected system for the distribution of electricity over a wide area, especially a network of high-tension cables and power stations.

Guarantee of origin (GO) – labels electricity from renewable sources to provide information to electricity customers on the source of their energy.

Heat network – a system for distributing heat generated in a centralised location through a system of insulated pipes for residential and commercial heating requirements such as space heating and water heating.

Heat pump, air source – a system which transfers heat from outside to inside a building, or vice versa.

Heat pump, ground source (or geothermal heat pump) – a central heating and/or cooling system that transfers heat to or from the ground.

Heat pump, water source – in an open-loop system (also called a groundwater heat pump), the secondary loop pumps natural water from a well or body of water into a heat exchanger inside the heat pump.

Horsepower – an old-fashioned unit of power equal to about 750 watts.

Hybrid vehicle – a vehicle using two different forms of power, such as an electric motor and an internal-combustion engine, or an electric motor with a battery and fuel cells for energy storage.

Hyperconsumerism – the consumption of goods for non-functional purposes and the associated significant pressure to consume those goods exerted by modern, capitalist society, as those goods shape one's identity.

Inflection point – an event that results in a significant change in the progress of a company, industry, sector, economy or geopolitical situation and can be considered a turning point after which a dramatic change, with either positive or negative results, is expected to result.

Infrastructure – the basic physical and organisational structures and facilities (e.g. buildings, roads, power supplies) needed for the operation of a society or enterprise.

Insulation – a material or substance that is used to stop heat, electricity or sound from going into or out of something.

Internal-combustion engine (ICE) – an ICE, such as the gasoline or diesel engine in your car, truck, tractor or bus, uses air as a working fluid, traps it inside the cylinders and adds heat by burning fuel.

Joule (J) – the SI unit of energy. About 100,000 joules are required to boil enough water for a mug of tea.

Kilo (K) – times one thousand, e.g. 1 kilowatt (1 kW) = 1,000 watts.

Kilowatt-hour (kWh) – a user-friendly unit of energy equal to 3.6 megajoules (MJ). One kWh is the energy delivered by one kW of power for one hour. Domestic electricity bills use the kWh as the basic unit of energy. The current cost of a kWh of electricity is around

14p and a kWh of gas costs around 4p. One kWh will boil enough water for about 36 mugs of tea and is roughly the energy contained in a dozen Hobnob biscuits.

Levelised cost of electricity (LCOE) – a concept used to compare the cost of energy coming from different renewable and non-renewable sources.

Light-emitting diode (LED) – a semiconductor light source that emits light when current flows through it.

Lithium – a soft, silver-white element of the alkali metal group that is the lightest metal known; it is used especially in alloys and glass.

Lithium-ion battery – a rechargeable battery that uses lithium ions as the primary component of its electrolyte.

Mega (M) – times 1 million, e.g. 1 megawatt (1 MW) = 1,000,000 watts.

Milli (m) – divided by 1,000, e.g. 1 milliwatt (1 mW) = 0.001 watts.

Mobility-as-a-service (MAAS) – a shift away from personally owned modes of transportation and towards mobility solutions that are consumed as a service.

Musk, Elon – technology entrepreneur, investor and engineer. He is the founder, CEO and lead designer of SpaceX; co-founder, CEO and product architect of Tesla, Inc.; co-founder of Neuralink; founder of The Boring Company; co-founder and initial co-chairman of OpenAI; and co-founder of PayPal.

Natural gas – a fuel consisting mostly of methane gas, extracted from under the ocean or ground.

Net metering – allows consumers who generate some or all of their own electricity to use that electricity any time, instead of when it is generated.

Nuclear fission – a nuclear reaction in which a heavy nucleus splits spontaneously or on impact with another particle, with the release of energy.

Nuclear fusion – a nuclear reaction in which atomic nuclei of low atomic number fuse to form a heavier nucleus with the release of energy.

Off-peak – a time when demand for something is less.

Oil – a viscous liquid derived from petroleum, especially for use as a fuel or lubricant.

Particulates – very small particles of a substance, especially those that are produced when fuel is burned.

Pescatarian – a person who does not eat meat but does eat fish.

Petrol – a petroleum-derived liquid mixture, primarily used as fuel in internal-combustion engines.

Plastic pollution – the accumulation of plastic objects in the Earth's environment that adversely affects wildlife, wildlife habitat and humans.

Plug-in hybrid electric vehicle (PHEV) – a type of hybrid electric vehicle that combines a gasoline or diesel engine with an electric motor and a large battery that can be recharged by plugging into an electrical outlet or charging station.

Pollution – the introduction of contaminants into the natural environment that cause adverse change. Pollution can take the form of chemical substances or energy, such as noise, heat or light.

Powertrain – all the components involved in moving a vehicle forward, including the battery, motor(s), power electronics and gearbox(es).

Public transport – transport of passengers by group travel systems available for use by the general public, typically managed on a schedule, operated on established routes, and charging a posted fee for each trip.

Range extender – a generator, usually powered by petrol, built into an electric vehicle; it may be used to charge the vehicle's battery while travelling in order to extend the vehicle's range. Unlike hybrid vehicles, range extenders do not provide power directly to the wheels.

Recycling – the process of converting waste materials to new materials and objects. It is an alternative to 'conventional' waste disposal that can save material and help lower greenhouse gas emissions.

Renewable energy – energy that is collected from renewable resources, which are naturally replenished on a human timescale, such as sunlight, wind, rain, tides, waves and geothermal heat.

Ride-sharing – any means of transportation in which multiple people use the same car, truck, van or other vehicle to arrive at a similar destination.

Science, Technology, Engineering and Mathematics (STEM) – a term used to group together these academic disciplines.

Seba, Tony – a world-renowned author, thought leader, speaker, educator and entrepreneur. He is the author of the Amazon number-one bestselling book *Clean Disruption of Energy and Transportation*.

SI – the international standard system of units including the metre, kilogram, ampere, volt, joule and watt. SI is an abbreviation of the French Système International.

Single-use plastics (or disposable plastics) – plastics that are used only once before they are thrown away or recycled. They include water bottles, straws and plastic bags.

Small modular reactor (SMR) – a type of nuclear fission reactor which is smaller than a conventional reactor; it is manufactured at a plant and brought to a site to be assembled.

Smart charging – the intelligent charging of EVs, where charging can be shifted based on grid loads and in accordance to the vehicle owner's needs.

Smart Export Guarantee (SEG) – expected to come into force in the UK on 1 January 2020, to replace the defunct feed-in tariff.

Smart meter – an electronic device that records consumption of electric energy and communicates the information to the electricity supplier for monitoring and billing.

Solar photovoltaic (solar PV) – a method of generating electricity by converting sunlight into solar energy using photovoltaic cells.

Solar thermal energy – a form of energy and a technology for harnessing solar energy to generate thermal energy or electrical energy for use in industry, and in the residential and commercial sectors.

Storage (or energy storage) – the capture of energy produced at one time for use at a later time. A device that stores energy is generally called a battery.

Subsidy (or government incentive) – a form of financial aid or support extended to an economic sector (or institution, business or individual) generally with the aim of promoting economic and social policy.

Sustainability – the process of conserving an ecological balance by avoiding depletion of natural resources.

Tera (T) – times 1 trillion, e.g. 1 terawatt (1 TW) = 1,000,000,000,000 watts.

Thermal storage – depending on the specific technology, it allows excess thermal energy to be stored and used hours, days or months later, at scales ranging from individual process, building, multi-user-building, district, town or region.

Thermostat – a component which measures the temperature of a room to allow a heating system to maintain the required temperature.

Tidal power (or tidal energy) – a form of hydropower that converts the energy obtained from tides into useful forms of power, mainly electricity.

Tipping point – the critical point in a situation, process or system beyond which a significant and often unstoppable effect or change takes place.

Transport as a service (or mobility as a service) – a shift away from personally owned modes of transportation and towards mobility solutions that are consumed as a service.

Veganism – the practice of abstaining from the use of animal products, particularly in diet, and an associated philosophy that rejects the commodity status of animals.

Vegetarianism – the practice of not eating meat or fish, especially for moral, religious or health reasons.

Vehicle to grid (V2G) – the energy concept whereby electric cars, and certain plug-in hybrid cars, could not only receive electricity from but send surplus electricity back into the electric grid.

Watt (W) – the SI unit of power. One watt equals one joule of energy per second.

Watt-hour (Wh) – one-thousandth of a kilowatt-hour. Sometimes used to express efficiency of electric vehicles, e.g. 250 Wh per mile. To change this to miles per kWh, divide 1,000 by the Wh per mile, e.g. 250 Wh per mile = 1,000 / 250 = 4 miles per kWh.

Watt-peak (Wp) – the peak power generated by a system under ideal conditions. The watt-peak value of a solar panel gives its electrical power output when it is bathed in one kilowatt of sunshine per square metre at 25°C, which is roughly the same as a clear sunny day. The output of wind turbines is given as their peak output rather than average output, which will depend on wind speeds.

Wave power – the capture of energy from wind waves to do useful work – for example, electricity generation, water desalination or pumping water.

Wind power, offshore – the use of airflow through wind turbines built in bodies of water – usually in the ocean, on the continental shelf – to harvest wind energy to generate electricity.

Wind power, onshore – the use of airflow through land-based wind turbines to provide the mechanical power to turn electrical generators, creating electrical power.

Zero carbon – achieving net zero carbon dioxide emissions by balancing carbon emissions with carbon removal or simply eliminating carbon emissions altogether.

Zero emission – refers to an engine, motor, process or other energy source that emits no waste products that pollute the environment or disrupt the climate.

Zero waste – a philosophy that encourages the redesign of resource life cycles so that all products are reused with the goal of no rubbish to be sent to landfills, incinerators or the ocean.

ABOUT THE AUTHORS

Pete Abson

National Grid, Public Affairs and Policy Senior Manager

Pete Abson is the public affairs and policy senior manager for National Grid. He has over 10 years' experience as a public affairs and policy professional, delivering successful campaigns and political engagement strategies for an FTSE 20 company.
@absonpeter

Monica Araya

Costa Rica Limpia, Founder & Costa Rica's EV Association, Co-Founder

Monica Araya works on consumer engagement in the shift towards renewable energy and zero-emissions electric mobility. She founded Costa Rica Limpia in 2014 (costaricalimpia.org) to bring people closer to clean technologies that improve everyday life. She also co-founded the Electric Mobility Association in Costa Rica. She works internationally, living in Costa Rica and Amsterdam. She writes frequently for international and local media, is a media commentator and has given talks in Latin America, Europe and Asia. Monica is an early advocate of the fossil-free Costa Rica vision, and her TED talk on this subject has over 1.2 million views.
@MonicaArayaTica

Roger Atkins

Electric Vehicles Outlook Ltd, Founder

Roger Atkins is a LinkedIn Top Voice for EV with a significant global following fast approaching 300,000 people. In and around the auto industry for over three decades, he's been sharing his electric vehicle narrative based on 15 years' experience in this amazing nascent industry. He's a relentless blogger, regular consultant and advisor to operators around the world, as well as a speaker and host at Clean Tech events.
@SidAtkins

Dr Kathryn Boast

University of Oxford, Outreach Officer

Kathryn Boast is a physicist and science communicator. Since completing her PhD on hunting for dark matter, she has been exploring exciting ways to share physics with the world, from interactive installations and colouring books to videos and stage shows. In particular, Kathryn believes in trying to open up science so that it is accessible for everyone.
@Kathryn_EB

Dan Caesar
Fully Charged, Managing
Director
With 25 years' experience in
business leadership, Dan Caesar
has invested the last 16 years
in the promotion of renewable
energy and electric vehicles,
and is deeply passionate
about tackling air pollution
and climate change. Clients
and employees have included
CORGI, Department of Energy &
Climate Change, Homebuilding
& Renovating Show, Wolseley,
Renewables Roadshow, Solar
Media, Accelerating Clean
Technologies and Ceres Power.
As *Fully Charged*'s managing
director, Dan is working with
Robert Llewellyn to take 'the
world's no.1 clean energy and
EV consumer channel' to a
bigger audience.
@FullyChargedDan

Tom Callow
BP Chargemaster, Director of
Communication and Strategy
Tom Callow has worked in
the EV and wider automotive
sector for 11 years. He formerly
chaired the motor industry's
EV group and has broad
knowledge of electric
vehicle products and
technologies. His role at
BP Chargemaster includes
engaging with the media, car
manufacturers, government
and other stakeholders,

and developing strategy.
@au_tom_otive

Dr Nicholas Carter
LG Electronics, Product
Training Manager
Dr Nicholas Carter is product
training manager with LG
Electronics, creating and
delivering training on energy
storage products. He has been
working in solar and energy
storage and driving electric for
over 17 years, currently sharing a
2011 Nissan Leaf and a 2017 Tesla
Model 3 with his wife, Georgia.
Nicholas previously spent three
years as technical training
manager at Tesla, introducing
Powerwall worldwide and
creating a complete training
program for installation and
service of Powerpack.
www.linkedin.com/in/nicholas-
carter-936b4313

Elinor Chalmers
Electric Vehicle Association
(EVA) Scotland, Director
Elinor Chalmers began her
electric dream in 2015 when she
bought her first electric vehicle,
a 24 kwh Nissan Leaf Acenta
Plus with striking blue alloys.
She shares her experiences
of owning an EV without a
home charger on social media.
Nissan customised an animal-
themed Leaf for her wedding
day at Edinburgh Zoo. In 2018
Elinor became the first female

director of the Electric Vehicle
Association (EVA) Scotland.
@She_sElectric

Graeme Cooper
Electric Vehicles, National Grid,
Project Director
Graeme Cooper leads National
Grid's work on electric vehicles/
decarbonisation of transport
by leading and coordinating
all the work relating to the
UK-regulated business of
National Grid. National Grid
separated the Transmission
Ownership (TO) from the
System Operation (SO) in April
2019, although since Graeme
has a corporate function, he
will be representing the TO
business of National Grid
externally. His work helps
the government, the energy
and automotive industry's
transition towards zero-
emissions vehicles. With over
a decade in communications
infrastructure and over a
decade in low/zero carbon
electricity, Graeme is a well-
known and highly respected
energy industry expert.
@bartlettsboy
www.linkedin.com/in/
graemecooper

David Hunt
Hyperion Executive Search,
CEO & Founder
David Hunt is a prominent
figure and thought leader

in the clean-energy sector. Hailed as a leading green entrepreneur by the *Financial Times*, David also presents at industry events such as EcoSummit, Energy Storage Europe and *Fully Charged* LIVE. He also hosts the podcast *This Week in Cleantech*. For his day job, David is CEO and founder of the leading cleantech executive search firm Hyperion Executive Search.
@davidHsearch

Phil Hurley

NIBE Energy Limited, Managing Director
Phil Hurley is Managing Director of NIBE Energy Limited, a subsidiary of NIBE Climate Solutions, with its headquarters in Sweden. NIBE is one of Europe's leading manufacturers in the domestic heating sector and Europe's leading heat pump manufacturer. NIBE offer high-tech solutions for heating, ventilation, cooling and heat recovery that reflect today's demand for sustainable construction. Phil has been involved in building services for over thirty years in various positions including contracting, design, product development, sales and marketing. He has been involved in renewable technologies for the last fifteen years and is a leading figure in promoting these technologies, in particular in the development of MCS, SAP Q, Benchmark and training courses in renewable technologies.
@NIBE_UK

Robert Llewellyn

Fully Charged, **Creator and Presenter**
Robert Llewellyn is a British actor, presenter and writer, famous for his thirty-year stretch as the rubber-masked mechanoid Kryten in the much-loved science fiction comedy *Red Dwarf*. With his interest in engineering, Robert then turned his hand to presenting the long-running TV gameshow *Scrapheap Challenge* and also *How Do They Do It?* and *Carpool*. In 2010, Robert launched *Fully Charged*, a YouTube show focusing on the future of electric vehicles and clean energy. *Fully Charged* has exceeded 55 million views around the world and Robert was recently acknowledged as 'Tech Legend' at the T3 Awards.
@bobbyllew

Dr Euan McTurk

Plug Life Television, Electrochemist
Dr Euan McTurk is an electrochemist and electric vehicle battery engineer who has been driving EVs since 2009. Euan completed his PhD in Materials Science at the University of Oxford, where he researched next-generation cell chemistries, before moving to WMG, where he developed methods to insert probes into cells from electric vehicles to determine how hard their performance can be pushed, how they fail and how to stop them failing. Now working as a battery electrochemist in Scotland, Euan launched *Plug Life Television* in 2018 to explain the fundamental workings of EV batteries in a way that anyone can understand, and to bust myths and misconceptions about electric vehicles.
@106Euan

Maddie Moate

Do You Know?, **Presenter**
Maddie Moate is a YouTube filmmaker and BAFTA-winning presenter, passionate about curiosity. She is the host of the BAFTA-nominated CBeebies series *Do You Know?*, BBC Earth's *Earth Unplugged* and CNBC's technology series *The Cloud Challenge*. One of the only family-focused 'edu-tubers' in the UK, Maddie has been creating educational online science content for the past seven years

and has amassed over 25 million views on her YouTube films.
@maddiemoate

Bjørn Nyland
YouTube, Blogger
Bjørn Nyland is a YouTube blogger who started making videos about Teslas. Over the past years this has grown into reviewing all other EVs as well. In the beginning he blogged as a hobby in his free time. In May 2017, he quit his office job and starting making videos full-time on YouTube. He is known for his road trips and real-world tests of EVs.
@BjornNyland

Emma Pinchbeck
RenewableUK, Deputy CEO
Emma Pinchbeck is an expert in renewable energy and decarbonisation. At RenewableUK, her team champions renewables, like wind, and the technologies that work with them, like storage, EVs or even hydrogen, working with governments and businesses in the UK and globally. She was previously Head of Climate Change at WWF-UK and holds advisory positions with the Energy Futures Lab and Energy and Physics Research Council. She is passionate about urgent action on climate change.
@Elpinchbeck

Tim Pollard
Pollard and Pollard, Principal
Tim Pollard is the principal of Pollard and Pollard, a consultancy service working in the fields of sustainability, energy and resource efficiency in the built environment. Tim was Head of Sustainability for Wolseley UK and was responsible for delivering the Sustainable Building Center at Leamington Spa as well as defining their renewables, energy and water-efficient products offer. In 2008 Tim was included in *Building* magazine's Top 40 list of 'Green Gurus' and was named as one of the West Midlands' Top 50 'Green Leaders'. Tim has also sat on the DECC RHI Industry Advisory Group, the HHIC Low Carbon Group and the NEF Advisory Council, and was chairman of the Water Using Products Group.
@LeamingtonSBC

Jonathan Porterfield
Eco Cars, Owner
Since 2011, Eco Cars has specialised in EVs (and PHEVs) and provided a personalised buying service to the trade and public in the UK, Republic of Ireland, Malta and France. Their mission is to change people's views of EVs and offer a trade buying service so that customers get the best possible value.

Mike Potter
DriveElectric, CEO
Mike Potter is the CEO of DriveElectric, leaders in zero-emission mobility and energy technology. They offer electric car and van leasing, charging equipment and advanced Crowd Charge vehicle to grid and energy technology.

Tim Shearer
Eco-Cars, Driver
Tim Shearer lives in Tain, Ross-shire, having relocated from the home of all things renewable: Orkney. He collects second-hand electric vehicles throughout the UK on behalf of Eco-Cars and delivers them to customers in Scotland. His longest trip was in a 40 kWh Nissan Leaf, from Bournemouth to Orkney. He drives a Kia Soul EV.
@TimsTravels1

Dr Nina Skorupska, CBE FEI
Renewable Energy Association, Chief Executive
Nina Skorupska is the chief executive of Renewable Energy Association (REA). Nina has over 30 years' experience in the energy industry, working in the UK, Germany and the Netherlands, and was the first female power station manager for RWE npower. She is a member of the board of Transport for London (TfL) and deputy chair of the

board of the WISE campaign 'to get a million women working in STEM industries by 2020'.
@NinaSkorupska

Jonny Smith
Presenter
Jonny Smith has been a motoring journalist, television presenter and online content maker since 1998. Starting out in print media on various classic, custom and new car magazines, he has now been on the presenting team of long-running Discovery series *Fifth Gear* for 17 series, as well as hosting and co-producing several car series for the BBC. Jonny has owned over 130 cars and back in 2012 began building a classic electric hot rod called 'Jonny's Flux Capacitor'. This humble 1974 Enfield 8000 car would become the world's quickest street legal EV – the record still standing today. Jonny is a firm favourite with *Fully Charged* fans, bringing his encyclopaedic car knowledge and technical curiosity to regular review episodes and weekly podcasts.
@Carpervert

Nadia Smith
Renewable Energy Association, Policy Analyst
Nadia Smith works across solar, energy storage, and electric vehicles policy. She supports businesses to understand and adjust to government policy, and communicates businesses' needs to government, to help the growth of the renewables industry. She has also worked in community energy, helping to set up and manage community-owned renewables projects, and is a director of South East London Community Energy Co-op.
@NadiaGreenNrg

Ben Sullins
Teslanomics, Data Geek
Ben Sullins' passion for sustainable transport – and Tesla's specifically – began when his son was born and he knew he wanted to have a safe, reliable vehicle that was also helping reduce CO_2 emissions. He began sharing his journey on YouTube by doing what he did for nearly twenty years in his career: finding the stories buried in data. This was a success online and has become Ben's full-time occupation. Ben plans to continue helping people realise the benefits of going electric until electric vehicles become the normal choice throughout the world.
@BenSullins

Angela Terry
One Home: Positive Solutions, CEO & Founder
Angela Terry founded One Home in 2018 to promote solutions to climate change. Angela has 20 years' experience in developing renewable projects, including pioneering community-owned renewable energy in the UK. She has contributed to government energy policy, worked on energy efficiency and worked as a carbon scientist for the forestry sector. She now liaises with the media to promote climate solutions and runs the UK's first one-stop-shop for climate action.
@angeterry

IMAGE CREDITS

Page 4 (1) © Gavin Rodgers /
Alamy Stock Photo; (2) © Valeriya Popova /
Alamy Stock Photo; (3) © xujun /
Shutterstock.com

Page 5 (4) © supergenijalac /
Shutterstock.com; (5) © Milesbeforeisleep /
Shutterstock.com

Page 6 (1) © meunierd / Shutterstock.com;
(2) © Ian Rutherford / Alamy Stock Photo

Page 7 (3) © nrqemi / Shutterstock.com;
(4) © Martyn Jandula / Shutterstock.com;
(5) © EQRoy / Shutterstock.com

Page 10 © Pictorial Press Ltd /
Alamy Stock Photo

Page 11 © WENN Rights Ltd /
Alamy Stock Photo

Page 13 © Allen Creative / Steve Allen /
Alamy Stock Photo

Page 22 © amophoto_au /
Shutterstock.com

Page 23 © Andy Morton /
Alamy Stock Photo

Pages 24–5 © Jenson / Shutterstock.com

Page 38 © Lorna Roberts /
Shutterstock.com

Page 44 © adrian arbib /
Alamy Stock Photo

Page 46 © myenergi

Page 63 © Zap-Map (zap-map.com)

Page 114 © P Cox / Alamy Stock Photo

Page 115 © Tramino / iStockphoto.com

Page 117 © Grzegorz Czapski /
Shutterstock.com

Page 120 © Kaukola Photography /
Shutterstock.com

Page 121 © VanderWolf Images /
Shutterstock.com

Pages 122–3 © Kaukola Photography /
Shutterstock.com

Page 130 © Tramino / iStockphoto.com

Page 131 © Grzegorz Czapski /
Shutterstock.com

Page 132–3 © Grzegorz Czapski /
Shutterstock.com

Pages 134-5 © MG Motor Ltd

Page 150 © Vauxhall Motors Limited

Page 153 © VDWI Automotive /
Alamy Stock Photo

Pages 154–5 © Grzegorz Czapski /
Shutterstock.com

Page 156 © Yauhen_D /
Shutterstock.com

Unbound is the world's first crowdfunding publisher, established in 2011.

We believe that wonderful things can happen when you clear a path for people who share a passion. That's why we've built a platform that brings together readers and authors to crowdfund books they believe in – and give fresh ideas that don't fit the traditional mould the chance they deserve.

This book is in your hands because readers made it possible. Everyone who pledged their support is listed below. Join them by visiting unbound.com and supporting a book today.

Niels Aagaard Nielsen
Anthony Abbot
Michael Abbott
Karl Adair
Ron Adolf
Dave Ahern
Michael Aikins
Mark Ainsworth
David Alcock
Keith Alcock
Stephen Aley
Carl Alldritt
Arlene Allen
Ashley Allen
David Allen
Emma Allen
Karen & Chris Allen
Lewis Allen
Stephen Allen
Tim Allen
Scott Allison
Keith Allwright
Gonçalo Almeida
Basil Alrabghi
Sören Anderberg
Hans Andersen
John Andersen
Brian Anderson
Moe Anderson

Bob Andersson
Daniel Andersson
Fredrik Andrén
Mark Andrew
Colin Andrews
Daniel Andrews
Martin Angus
Jeremy Anson
Mark Antrobus
Steven Archibald
Jo Argent
Bill Armstrong
James Armstrong
Thomas Arntzen
Adithya Arun
Bruce Asbestos
Martin Ashby
Steven Ashford
P. Joel Assamoi
Jay Atkinson
Nick Aurelius-Haddock
David Austin
Gregor Autischer
Robert Avery
Chris Avis
William Ayling
B & R
Robin Babicek
Captain Badbeard

Stefan Baer
Richard Bagnall
Kov Bahadori
Diana Bailey
Greg Bailey
John Bailey
Ray Bailey
Kevin Baillie
Paul Bain
Rob Baird
Allan Bairstow
Robert Baker
William Baker
Michael Ball
Mike Ball
Jason Ballinger
Martin Barlow
Mark Barnfield
Mike Barratt
Teo Barrault
Simon Barrick
Mark Barrington
Doug Barth
Lewis Barthaud
Matthew Barton
Jacob Bartynski
James Bassett
Chris Basterfield
Marko Baštovanović

Matthew Bate
Terrence R. Bauman
Tilman Baumann
James Baxandall
Jamie Bayliss
Edward Bayly
Andrew Beale
Matt Beard
Kevin Beason
Graham Beaumont
Steve Beaumont
Peter Beckman
Parveen Begum
John Beisley
Adrian Belcher
David Bell
James Bell
Patrick Bendure
Tim Benford
Laurie Bennett
RobertJB001 Bennett
Ricky Bennici
David Benoy
Rodney Bento
André Berger
Ben Berger
Claes Bergman
Kelley Bergmann
Jaap Berkers

Christian & Peter Bernhofer
Trevor Berridge
Garth Berry
Graham Berry
Martyn Berry
Michelle Bertelsen
George Bethell
Steve Bickle
Jacob Biddulph
Daniel Biggins
David Biley
Jack Biltcliffe
Richard Bingham
The Bionic Cyclist
Malcolm Bird
Stephen Bird
Richard Birkett
Gordon 'Solar Homer
    Simpson' Bishoff
Mike Bishop
Per Bitsch Larsen
Tim Bjerg
Simon Blackmore
Simon Blackwood
Christopher Blake
Jamie Blake
John Blanckaert
Chris Blanksby
Michel Blier
Bluesmith
Steve & Trudy Boardman
BobHW PassWithBob.com
Anthony Boden
Caro Bois
Christof 'Zottel' Bojanowski
David N M Bond
John Bond
John Bone
Roy Bone
Adrian Bonet
Marco Booth
Sanjeeva Borahalli
Iain Borden
Robert Borley
Steinar Børve

Ian Boulstridge
Alexander Bourgett
Anthony Bourn
Ayrton Bourn
Mathew Bowden
Phillip Boys
Kieren Bradley
Paul Bradley
Alan Bradshaw
John Braines
Mat Braithwaite
Peter Brazier
Nicholas Brealey
Mark Breese
Mark Brennan
Raymond Brennan
Patrick Breuer
Nick Briars
Stephen Brigdale
Christof Brill
Henry Briscoe
David Britton
James Broady
Daniel James Brogan
Dawn Marie Joanne Brogan
Alan & Angie Brook
Stephen Brook
Elliot Brooke-Freeman
Jon Brooks
Wilfred Brooks
Marcel Brouwers
Ian Browb
Andrew Brown
Chris Brown
David Brown
Kevin Brown
Matthew Brown
Neil Brown
Samuel Brown
Roger Browne
William Brownlie
John Bryan
Stephen Buckley
Peter Bull
Michael Bullock

Mark Bulpitt
Jens Büntjen
Marco Burattin
Jaap Burger
Thomas Burghout
Matt Burke
Andreas Burmberger
Christine Burns
Tony Burns
Sam Burnstone
Christopher Burrows
Stephen Busby
Alison Bushnell
Anthony Butcher
Benjamin Butler
Mark Butterworth
David Bye
Ian C.
Con Cahill
Jeffrey Cairns
Joe Calabrese
David Calder
Alexander Calvey
Joseph Camacci
Wayne Campbell
The Campbell family
Simon Canfer
Martin Cantwell
Matthew Canty
Michael Carden
Joel Carey
Paul Carlson
Tim Carnegie
Maurice Carney
Joseph Carr
Anthony Carrick
Andrew Carruthers
Gregory Carter
Henry Carter
Joe Carter
Phil Carter
Rod Carter
Stephen Robert Carter
Stuart Carter
Paul Cartwright

Brian Casselman
Tony Castley
Mike Caulfield
Chris Cavanagh
Aryan Chadha
Joshua Chadney
Gordon Chaffin
Edward Chambers
David Chandler
James Chapman
Olly Chapman
John Andrew Charles
Adam Chattington
Chris Chatwin
Simon Cherry
Nina Chiu
Anato Chowdhury
John Christen
Juan Christian
Michael Christie
Tom Christopher
Mieszko Cichowicz
Keith Cirkel
Andrea Ćirlićová
Octav Ciuca
Richard Clack
Nick Clare
James Clark
Jasper Clark
Paul Clark
Stuart Clark
Andy Clarke
Chris Clarke
Damian Clarke
Jenny Clarke
Sam Clarke
Steve Clarke
Stuart Clarke
Robin Clay
Alan Cleaver
Richmond Clements
Peter Clewes
Jason Clifford
Matt Coates
Graeme Cobb

Nick Cohn
Matthew Colbourn
Jack Coldridge
Hamish Cole
Paul Cole
Phil Cole
Andrew Coleman
James Collett
Graham Colley
Francis Collier
Tristan Collier
Christopher Collins
Paul Collins
Paul Compton
Paul Conway
Pól Conway (Bukeejit)
Kevin Conyers
Carl Cook
Jim Cook
David Cooke
Jeremy Cooke
John Coombes
Sean Coomey
Steven Cooney
Anthony Cooper
David Cooper
John Cooper
Mark E Cooper
Paul Cooper
Nick Coray
Tina Cordon
Simon Corlett
Joseph Cormack
Trenna Cormack
Robbert Cornelissen
Jonathan Corney
Nigel Cornwall
Quinn Corrigan
Edwyn Corteen
David Cosker
Stephen Cosser
Radu-Iulian Costin
Alan Cottage
James Cotterell
Dan Coulcher

Patrick Count
Stephen Courtnadge
Pete Cousins
Stephen Cowans
Andrew Cowell
Stuart Cragg
Ollie Cramer
Jon Cranefield
Brent Crawley
Remie Cremers
Robert Crichton
Nigel Critten
Daniel Croft
Nigel Crompton
Matt Croucher
Ian Crouchman
Israel Crowe
Marc Crowley
Michael Crump
Mike Cuchna
Christian Cuddington
Egils Cukermanis
Paul Cummings
Peter Cunningham
Ian Curley
Joy Curran
Alasdair Currie
Tim Curtis
Roger Cuthbert
Liz Cutter
Brian D. & Heather A. Smith
Maarten Leo Daalder
Gytis Dagilis
Grant Daly
Mick Dann
Mark Darlington
Tom Dauben
Claudio David
Andrew Davidge
Graham Davidson
Gareth Davies
Jeremy Davies
Nathan Davies
Rhys Davies
Zoe Davies, Lee Johnson

Gary Davison
Peter Dawes
Dave Dawson
Karl Dawson
Martin Day
Maikel De Bakker
Dean De Blieck
Christopher de Chazal
Jean-Paul de Chazal
Rutger De Croon
Sam de Freyssinet
John de Klerk
Marcel de Pender
Brent De Saegher
Ard de Zeeuw
Graham Dean
Rob and Robin Dean
    Western Australia
James Dearman
Martin Dell
Paul Denney
David Denning
Brendan Dennis
Bernd Derksen
Gareth Desborough
Ronan Devitt
Scott Devlin
Ellen Dexter
Daniele Di Carlo
Martin Dibben
Ken Dickie
Richard Dickins
Mark Dickson
Matt Dixon
Steve Dobson (Dobbo)
Nick Dogramadzi
Christian Dolansky
Mark Dommett
Teodor Donici
Stuart Dore
Joachim Dorn
Jeff Dowling
Rob Dowson
Andy Draper
Max Drennan

Peter Duckmanton
John Dudney
Heather Duff
Neil Duff
Clive Duffy
Mark Duffy
Alex Dufournet
Sandra Duggan
Dr Douglas Duncan
Sebastian Duncan
Christopher Dunn
George Dunnett
David Dupplaw
Mark Durbin
Jillian Durham
Oliver Dutton
Paul Dutton
Andy Duval
Phil Dyson
Ian Eagland
Chris Eason
Ryan Eby
Kim Edwards
Matthew Edwards
Scott Edy
Nan Eelman
Matthew Eglin
Gordon Elliott
Michael Elliott
Ben Ellis
Kevin Ellis
Matthew Ellis
Peter Ellis
Rob Ellis
Stephen Ellis
Steve Ellis
Mark Elvin
David Emeny
Enigmabob
G. Erlendsson – ENSO
European Marine Energy
    Centre (EMEC)
EV Digest Ltd evdigest.co.uk
Aron Evans
Kate Evans

Luke Evans
Rob Evans
Tim Evans
Carsten F.
Hervé Fache
Daniel Fairbairn
Mark Fairbairn
Paul Fairless
Jonathan Fallman
Lee Farman
Chris Farmer
Richard Farmer
Adrienne Farrow
Gregory Fattorini
Dean Faulk
Anthony Featherstone
Jessica Feinleib
Duncan Ferguson
Jonathan Ferguson
Miha Ferlež
Benjamin Fernandes
Jacinto Leao Fernandes
Andrew Ferrier
Mark Fiddament
Peter Fido
Mark Field
Dean Fielding – Keystone
    Future Homes
Peter Fieldsend
David Finch
Roger Finch
Stephen Finch
Gregor Findlay
John Finlay
Lance Finney
Patrick Fischer
Edward Fisher
Michael Fisher
Peter Fisher
Sean Fitzgerald
Paul Fitzpatrick
Bård Fjukstad
Tony Flahive
Clive Flatau
Andy Fletcher

Julian G Fletcher
Justin Fletcher
Jacques Flibotte
David Flood
Patrici Flores
John Foden
Mike Foden
Graham Forbes
Keith Forster
Robin A Foster
Chris Fox
Nick Fox
Rachael Fox
Pedro Fradique
Jamie Francis
Martin Fraser
Loz Freeman
Richard French
John Frewin
Urban Friberg
Adam Frost
Shawn Fry
Mark Fulbrook
Steve Fuller
furriephillips
Barry Fussey
Andrew Gaffney
Blair Gallagher
Conor Gallagher
Mark Gamble
Paul Gambrell
Joseph Gander
Sandy Gardner
Bernie Garland
Mark Garnett (Worcester)
Gary Garytodd
Lucas Gasenzer
Douglas Gault
Helen Gavin
Jonathan Gay
GDad (Robert W Aldis)
Liam Geary
Joshua Geisinger
R George
Ruud Gerritsen

Erik Geurts
Anil Ghatikar
Matthew Gibbens
Dominique Gibeau
Terry Gibson
Chris Gilbert
Shaun Gilbert
John Gilburt
John Gill
Ben Gillam
James Gilles
Ewan Gimson
Dave Ginnane
Richard Gledhill
Lee Glenister
Philip Godfrey
Edward Godsell
Michael Gomes
Emmanuel Gomez
Roy Goode
Lewis Goodell
Wayland Goodliffe
Karl Goodloe
Paul Goodridge
Colin Goodwin
Heide Goody
Mattias Göransson
Sean Gordon
Raymond Gore
Richard Gorman
James Gorrill
Pete Gorton
Vadim Goryunov
Graham Goss
Rob Gowman
Alex Graham
Mark Graham
Nicolas Grant
Richard Grant
Peter Grape
Kieran Gray
Andrew Green
Martin Green
Ben Greenberg
James Greene

Martin Greenwood
Paul Gregg
Edward Gregson
Daniel Grenyer
Jeremias Grenzebach
Kim Greve
Stephen Grier
Francis Griffin
Ian Griffiths
Paul Griffiths
Ragnar Már Grinde
Grumstrup
Chris Gryce
Simon Gueissaz
John Gunnee
Samuel Guss
Romain Guyony
Des Gwinnell
Jamie Gwozdzicki
Neil & Stacey Habergham-
    Mawson
Simon Hackett
Bill Haddock
Jamie Hailstone
Malcolm Haim
Chris Hales
Sam Hales
Andy Hall
Chris Hall
Michael Hall
David & Karin
    Halliday
Max Halliday
James Hammond
William Hancock
Geoff Hancox
Stuart Hannah
Alex Hansford
Brian Charles Hanson
Ian Hanson
Mark and Alison Hanvey
Eamon Harbison
Lisa Harcombe-Minson
James Hardie
Andrew Harding

David Harding
Filip Harding
Chris Hardisty
Mike Hardy
Richard Harkness
Steve Harper
Dr. Steven Harris
Matt Harris
Pelle Harris Krog
Bruce Harrison
James-Ross Harrison
Stuart Harrison
Ethan Harrold
Barrie Hart
Tessie Hartjes
Martin Harvey
Corentin Hatte
Tim Hawker
David R Hazlehurst
Ian Heath
John Heaton
Gustav Hedelin
Lewis Hedges
Karl Helliwell
Nigel Henretty
Jon Heras
Martin Herbert
Andreas Hermansson
Axel Hermansson
Julian G Herring
Andrew Hewitt
Jim Hewlett
Frankie Heyzak
Garin Hiebert
Paul Higgins
Stu, Katherine & Izzy
    Higgins
Ashley Hill
Ben Hill
Shaun Hill
Brian Hillan
Tyler Hilliard
Andrew Hills
Lahn Hinchliffe
Robert Hine

Kasper Hjorth
Robert Hoare
Ray Hobby
Sean Hockett
John Hodder
Alex Hodgson-Bamford
Bradley Hogan
Ollie Hogan
Fred Hogendorn
David Hogg
Gregory F. Hogg
Jason Holdcroft-Long
Stephen Holden
Andy Holland
James Holmes
John Holmes
Alex Hong
Thomas Hooper
Gareth Hope
Gareth Horne
Tim Horrox
Neil Horsley
Martin Horswood
Tim Hoskin
Matthew Houghton
Simon Houghton
Gerard Houwing
Bill Howard
John Howard
Paul Howarth
Andrew Howe
Alan Howie
Mark Hoyle
Glyn Hudson
Graham Hudson
Adam Huet
Paul Huggins
Tom Hughes
Alex Hulley
Paul Humphrey
David Hunt
Darren Hunter
Phil Hurley
Pete Hurst
Michael Hutchinson

David Hutton-Potts
Christopher Iddon
Mark Iliff
Ian Ingham
Alexander Innes
Soames Inscker
Ion DNA
Martin Isherwood
Hisato and Christine Ishii
Jamie Ivory
Peter Jackman
Carl Jackson
Jeremy Jackson
Paul Jackson
Will Jacobs
Alexander Jaeger
Martin James
Larry Jaques
Klaus Jaritz
Will Jarman
James Jean-Louis
Colin Jeffrey
Mike Jelfs
Dan Jenkins
Nathan Jenkins
Paul Jenkinson
Martyn Jennings
Henrik Jensen
Alexis Jenson
Ralph Jenson
Johan Jigsved
Denny John
Gutierrez Johnny
Adam Johnson
Mark Johnson
Natalie Johnson
Ben Johnston
James Johnston
Paul Johnston
Pete Jolliffe
Per Jonasson
Andrew Jones
Erin Jones
Jack Jones
Mark Jones

Peter Jones
Phil & Derek Jones
Rob Jones
Roger Jones
Rebecca Josefsson
Ian Joseph
Andrew Josey
John Joyce
Marçal Juan Llaó
Robin Julian
Steve Julien
Parmi Jutlay
Nicolai Kaasgaard Bonde
Swaroop Kagli
Peter Kamei
Martin Kamermans
Rob Kamp
Zdenko Kastler
Michael Kaufman
Emma Kavanagh
Mihaly Kavasi
Mark Keane
Simon Keane
James Keeble
Kieran Keegan
Daniel Keighobadi
Timo Kellenberger
Matthew Kelly
Adam Kemp
Ella Kennedy
Nick Kennedy
Mark Kennedy Lewis
Paul Kent
Aleksander Kenton
Paul Kenyon
Gavin Keogh
Ian Kerr
Ian Kew
Lazarus Kidd
Dan Kieran
Bradley Kieser
George Killaspy
Boyd Kilsby
John A G Kime
Adam Kingsley-Williams

Tim Kingswell
Jerry Kingzett
Jesse Kinross-Smith
Luke Kirby
Bhupinder Klair
Ian Knight
Jon Knight
Hendrik Kommerie
Mateusz Kot
David Kotila
Maximilian Kraus
Sebastian Krause
Troels Kristensen
Petr Křivonožka
Marc Kudling
Klaus-Otto Kuennemeyer
Wojciech Kurlapski
Dan Kyle-Spearman
Stephen Laing
Trevor Laley
Hutchison Lam
Piotr Lamacz
Bradley Lamb
Eugene Lambert
Gerry Langford
David Larder
Mark Larsen
Per Larsson
Tom Last
Steve Laughlin
Jon Launder
Karo Launonen
Peter Lawrance
Mike Lawrence
Phillip Lawrence
Nigel Lax
Alison Layland
Stefan Lazic
Philip Le Roux
Daniel Leahy
Jonathan Leake
David Leale
Dave Lee
Lawrence Lee
Jason Legg

Adam Leggo
Matt Lemon
Richard Lenderyou
Graham Lennard
Rachel Lenton-Leigh
Iain Lettice
Colin Leuthold
Colin Lev
John Levy
David Lewis
Gareth Lewis
Jon Lewis
Ladonna Lewis
Mark Lewis
Paul Lewis
Rhodri Lewis
Rhydian Lewis
Belinda Lewsley
Iain Liddle
Adrian Liechti
Pieter Lieverse
Zach Lindsey
Martin Linklater
Sebastian Liske
David Llewellyn
David Lloyd
Ronnie Lloyd
Tom Lloyd
Dave Lockhart
Demetri Loizou
Bob (Bobby Bent Teeth)
   Lomax
Derek Long
Alwin Look Ken Ming
Maurice Lopez
Ed Lord
Sean Louttit-Henighen
Claiton Lovato
Lubelhu
Finbarr Lucas
Jay Lucas
James Luck
James Lucy
Jordon Ludlam
Peter Lugerbauer

Mark Lund
Mikael Lundgren
David Lydford
Colin Lynch
Philip Lynch
Merlin Lyons
Darren Lyth
John Mac Canna
Conor Mac Fadden
Russell Macdonald
Ewan MacGregor
Alexander Mächler
Doug Mackay
Alec MacLean
Benjamin MacLean
Neil Maddox
Shannon Madigan
Stephen Madley
Arpad Madocsai
Anders Madsen
Jennifer Magill
Paul Magor
Kyle Mahan
Michael Mahon
Tony Major
William Makant
Steven Malatesta
Matthew Mammana
Marcus Mangelsdorf
Oliver Mankowski
Nick Manzi
Winfried Markert
Sascha Markham
Richard Marlow
Gill & Tony Marsh
Patrick Marshall
Stephen Marshall
Jonathon Martens
Hampus Mårtensson
Alastair Martin
David Martin
Graham Martin
John E P Martin
Matt Martin
Paul Martin

Dominik Martinicky
Nick Mason
Gianluca Massera
Dave Masters
Brett Masuoka
Anastas Mateev
Ján Mátik
Gary Matsell
Richard Mawby
Richard Mawle
Jeff May
Chris Mayman
Benjamin Mayne
Daniel McAdam
Don McAllister
Gavin McAllister
Francis McCabe
Declan McCaffrey
John McCann
Stephen McCarthy
Bernard McCarty
Phil McCormick
Paul McCullough
David McDermott
Andy McDonald
Mark McDonald
Peter McGinn
Rab T McGregor
David McGruther
John McGuinness
Craig McHugh
Iain McIntyre
Malcolm McIvor
Rob McKay
Andrew McKeown-Henshall
Ross McLachlan
Andi McLean
Michael McLean
Martin McMahon
Kevin McNally
John Meaking
Svein Medhus
Chris Meilak
Jonathan Melhuish (Verkla)
Neil Mellor

Mark Melocco
Ole-Christer Melvold
Roland Mercatoris
Susan Metcalfe
Tracey Meyer
Paul Middlicott
Wolfgang Mika
Nigel Miles
Don Miller
Gary Miller
John Miller
Tony Millington
Graham Millman
Scott Millns
Lee Mills
Marc J Milne
Nigel Milne
Charlie Milner
Robert Mines
David Minett
Dominic Minett
Howard Mitchell
Paul Mitchell
Stephen Mitchell
John Mitchinson
Jorma Moksi
Thibault Molleman
Simon Moloney
David Monks
Nick Monypenny
Simon Moore
John Moores
David Morgan
Paul Morgan
Robbie Morgan
Steve Morgan
Ian Morris
Neil Morris
Nigel Morris
Russell Morris
Stephen Morris
Daryl Morton
Michael Mosedale
Chris Mountain
Philip Mousley

Tom Mowlam
Mark Muehl
Neil Muggleton
Robert Muir
Michael Mulloy
Tom Mumford
Robert Munro
Bob Murphy
Max Mustermann
CJ Myers
Richard Myers
Andrew Naldrett
Pranay Nangia
Roger Narten
Richard Nash
Carlo Navato
Tina Nawrocki
Graham Naylor
Wesley Neal
Chris Neale
Lukas Neckermann
Alex Neill
Dan Nelson
Corin Nelson-Smith
Gonçalo Neves
Jacob Newman
Jim Newman
Will Newton
Jon Niccolls
John (sol) Nicholls
Richard Nicholls
David Nichols
Colin Nicholson
Duncan Nicol
Janwillem Nicolai
Andrée Nienkerk
NIEVO.org
Jan Noben
David Noden-Hooper
Tony Norfolk
Joakim Norldander
Lorenzo Novoa
Chie Nwawudu
Maxymilian Nytko
Chris O'Brien

Ian O'Brien
Eoin O'Connell
Victoria O'Leary
Derek O'Neil
Tom Obdam
Wojciech Ogrodowczyk
Gunnar René Øie
Thomas OKeefe
Matthew Oldfield
Simon Olewicz
Pieter O. Olivier
Lars Olivius
Jonas Öman
Frank Op 't Veld
Tony Orchard
Igor & Tatyana Orlovich
Pete Osborn
Luke Osborne
Jesper Østergaard
Outback Tesla
Jamie Owen
Robbie Owen
Edward Ozgowicz
Alasdair Page
Jonathan Pallant
Colin Palmer
Jason Palmer
Andrew Pam
Brian Panico
Anastasios Panousakis
Mark Pantry
Martyn Paradise
Kevin Parichan
David Parker
Kevin Parker
Ross Parker
Tim Parkes
Michael Parkinson
Daniel Parnham
Stephanie Parratt
Aurimas Parsonis
Martyn Parsons
Mike Parsons
David Partington
Niels Partoft

Greg Partridge
Neil Partridge
Richard Pascoe
Rajan Patel
Tom Paterson
Simon Patmore
Carl D. Patterson
Chet Pattni
Mattias Påvall
Seiki Payne
Ben Paynter
Jimmy Pea
Trevor Peabody
Lewis Pearson
Phil Pearson
Lesley Peere
Ander Pelayo
Niall Pembery
Pam Penkman
Nigel Pentland
Carlos Perez
Andrew Perry
Darrell Perry
Mikael Persson
Christian Peter
Krivansky Peter
Dan Peters
Paul Peterson
Gaëtan Philippot
Tim Phillips
Lewis Philpott
Roland Pickering
Matthew Pickett
Lance Pickup
Johan Pieterse
Ian Piggott
Richard Piper
Joost Pisters
Simon Pitkin
Roo Pitt
Tony Pitt
Martin Platts
John Pluck
The PlugSeeker
Stuart Pocklington

Justin Pollard
Polly, Nomio & Bridget
Nick Pont
Robert Pool
Roger Pope
David F Porter
Johan Posdijk
Michelle Potter
Mike Potter
Steve Potter
Sean Powell
Andrew Michael Powers
John Powles
Ravi Pradhan
Allen Pratt
Philippe Preaudat
Martin Preston
David Price
Jonathan Price
Luke Price
Richard Price
Simon Price
Richard Proctor
Christopher Pryor
Craig Pugsley
Chris Pullen
Adam J Purcell
Derek Purnell
Marc Qualie
Adrian Quinn
Andrew J Quinn
Kevin Quinn
Seosamh and Saoirse Quinn
Ashir Qureshi
Timothy Raffler
Silvio Rahm
Stephen Rail
Dave Ramsbottom
Christopher Ramsey
Sarah Rapley
Quentin Rasmont
Søren Bartels Rasmussen
Steven Rattray
Ville-Veikko Raty
Jim Ray

Haitham Razagui
Charlie Rea
James Read
Colette Reap
Simon Reap
Mark Reason
Tony Reed
Allan Rees
Ciaran Regan
Dieter Rehbein
Tobias Rehm
Donnie Reid
Simon Reid
Brendan Reidenbach
Craig Reilly
Derek Reilly
Adrian Reith
Robert Rennie
Julian Rex
Brian Rhodefer
John Rhodes
Frazer Richards
James Ricketts
Shachar Ricklis
Paul Ridings
Paul Ridley
Rollo Rigby
Nick Riley
Xavier Riley
Gareth Rimmer
John Ritchie
Dan Rivett
Andy Rixon
Alessandro Rizzo
Joshua Roache
Bruce James Roberts
Darren Roberts
David Roberts
Neil E Roberts
Peter Roberts
Peter James Roberts
Phil Roberts
Matthew Roberts
    (Matthew1471)
Alan Robertson

James Robinson
Paul Robinson
Rachael Robinson
Alex Rocha Ura
Mike Rodger
Jeremy Roebuck
Daan Roest
Brad Rogers
Michael Rogers
Christopher Rollin
Jason Rollins
David Rollinson
Jonas Roothans
Ken Ross
Sarah Gail Ross
Robert Roughton
Eric Rouviere
Eddie Rowe
Christopher Rowlands
Justin Rowlands
Walter Rowntree
Joe Roza
Daan Rozemeijer
Daniel Ruch
Daniel Ruff
Felix Ruponen
John Rush
Dionne Ryan
David Sailer
John Sammons
Peter Sammut
Mark Sanders
    (Lotus_Elon@77A)
Tom Sansome
Ramin Saraby
Jurgen Sasse
Steve Saul
Derek Saunders
Jason Savage
Giovanni Savastano
Paul Sawyer
Manuel Schallar
Hans Schenkelaars
Florian Schifferdecker
Finn Schmidtke

Michel Schmitt
Christian Schneider
Fabian Schneider
Jens Schneider
Tom M.H.W. Scholle
Mikael Schönning
Mike Schooling
Bertil Schou
Manuel Schröder
Thorsten Schüler
Kim Scott
Peter Scott
Victor Seal
Daniel Seifert
Kjetil Seljelid
Thomas Setchell
Colin Sexton
Scott Sexton
Andrew Seyes
Arif Shabbir
Siddharth Shah
Graham Shapcott
David Sharp
Tim Sharrock
Calum Shaw
Mike Shaw
Rob Shaw (RSThinks)
Ray Sheehan
Darran Shepherd
Terence Shepherd
Graham Sherlock
Karl Sherratt
Alan Shields
Arron Shields
Richard Shilling
Jonathan Shine
James Shinners
Jason Shore
Melanie Shufflebotham
John Shuttleworth
Rich Sibbick
Jonas Sidaravicius
Magdalena Siepka
Mitch Siepka
Bart Sigger

Jamie Siggers
Andrew Sillwood
Ruudi Silmann
Andy Silvester
Michael Simpkins
Kevin Simpson
Bob Sincick
Anthony Sinclair
Sebastian Sjöberg
Ben Skillen
Iolo Slaymaker
IJsbrand Slob
Harry Smale
Alan Smith
Corey Smith
Daniel Smith
Darren Smith
Howard Anthony Smith
Lael Smith
Martin Smith
Michael Smith
Nic Smith
Steve Smith
Bruno E.M. Smulders
Curt Snyder
Joe Sobkowiak
Wojtek Socha
Raphael Sofair
John Softley
Mike Sollom
Pedro Sottomayor
Darren Souter
Richard & Lil Souter
Alexander SpannerMonkey
Peter Spark
Greg Spedding
Loe Spee
Trevor Speering
Rob Spencer
Stephen Spencer
Peter Spicer-Wensley
Tim Spielman Jr.
Andy Spiers
Lubo Splino
Tobias Sprenger

Steve Springett
Chris Squires
Mel St Pier
Peter Stace
Tom Stacey
Tony Stamford
Geoff Staniforth
Keir Stanley
Tom Steadman
Zach Stednick
Oliver Stepanovic
Andrew Stephens
Ed Stephens
Matt Stephens-Rich
Sam Stephenson
Fredrik Sterngren
Chris Stevens
Adam Stewart
Brian Stewart
Duncan Stewart
Marcus Stewart
Ole Stobbe
Daniel Stockwell
Carl Stone
Dave Storey
James Strack
Howard Strickland
Justin Strik
Les Strong
Mark Stubbs
Martin Stubbs
Phil Stubbs-Thomas
Thierry Stucker
Andrew Stuttard
Andy Stvan
Gareth Suddes
Richard Paul Sullivan
Richard Summers
David Sunderland
Ed Svoboda
Dan Swain
Dennis Sweet
Jonathan Sykes
Matthew Sylvester
Daniel Symes

Jason Symonds
Michał Szade
Thomas Szoldatits
Dan Tamone
Paul Tang
paul_tanner Tanner
Robin Tanner
Paul Targett
Darren Tartaglia
Andy Taylor
Frank Taylor
Graham Taylor
Guy Taylor
Kevin Taylor
Mark Taylor
Michael Taylor
Richard Paul Taylor
Roger Taylor
Robert Telford
Simon J Telford
Paul Temple
Bohdan Terlecky
Richard Ternouth
Robert Terry
Tesla Owners Club Western
   Australia
Tony Thick
Heiner Thielmeier
Chris & Michelle Thomas
Cunningham Thomas
David Thomas
Derek Thomas
Mark Thomas
Philip Thomas
Jay Andrew Shaw
   Thompson
John Thompson
Mark Thorndyke
Adam Thornton
Craig Thornton
Trevor Thorogood
Anders Thorup
Steven Tibbs
Stephen Tidy
Ben Tilley

Jon Timmis
Roger Tipper
John Tisbury
Daniel Toman
Giulio Tonellato
Stefano Tonelli
David Tones
Phil Tordoff
Grant Totten
Roy Tovey
Ole Traumüller
Nick Traynor
Graham Triggs
Paulo Trigo
Achim Trumpfheller
Red R R Tuby
Simon Tucker
Iain Tunmore
James Tunnicliffe
Peter Turner
Mathew Tuttelman
Alistair Tuxworth
Ian Tyrrell
Lee Tyrrell
Roger Tyrrell
Darryl Underhill
Herbert Unger
Andrew Urquhart
David Urquhart
Bjørn Utgård
Tom Van Aardt
Oliver van Bilsen
Maarten van Casteren
Sander van de Ven
Stefan van den Berg
Nicolas van den Cloot
Rob van der Wouw
John P van Dieken
Linus Van Hove
Mark van Keulen
René Van Landeghem
BPvL van Leeuwen
Pim Fabian van Luik
Bart van Raalte
RAJ van Riel – LFP.support

Franky Van Rietvelde
Hendrik van Triest
Jeffrey van Vegten
Bart van Wijk Grobben
Salim Vanak
Stephen Vard
Tamas Varga
Benjamin Verazzi
Andrew Verney
Bert Verschuren
Hugo Vicente
Derek Vickers
Paul Villeneuve
Alex Vindel
Peter Viner-Brown
Daniel Vogedes
Koen Volkers
Ethan Vonderheide
Kuba Waliński
Antony Walker
Mark Walker
Peter Walker
Matthew Wallace
Richard Waller
Josh Wallis
Russ Walliss
Nick Walpole
Paul Walters
Steven Walters
Cody Ward
David Leslie Ward
Mark Ward
Michael Ward
Mike Ward
Steve Ward
Chris Warner
Richard Warner

Jack Warren
John Warren
Norman Warren
Michael Warwick-Sanders
Chris Wass
Mark Waters
Richard Waters
Steven Lloyd Watkin
Chris Watkins
Ian Watkins
Malcolm Watkinson
Kit Watson
Scott Watson
Vivian Watts
Andrew Waugh
Andrew Webb
Peter Webber
Stephen Weblin
Lola & Jude Webster
Mark Webster
Sam Webster
Oliver Weeks
Lennart van der Weide
Mauritz Weins
Thomas Weiss
Steve Wellington
David Wells
Karen Wells
Marc Welter
Mark West
Tom West
Duncan Westland
Phill Wheatley
Matt Wheeler
Paul Whelan
Chris White
David White

Gerry White
Matt White
Robert White
Brian Whitehead
Kevin Whitehead
Rob Whitney
Garry Whittaker
Ben Whittle
Brian Whittle
James David Whyte
Warren Whyte
Nigel Wickenden
Charles Wilcockson
James Wilford
David Wilkins
Brian Wilkinson
Richard Wilkinson &
    George Wilkinson
Jon Wilks
Pieter Willems
Ceri Williams
Christopher Williams
Jeremy Williams
Matthew Williams
James Williamson
Ray Williamson
Jacob Wills
Craig Wilson
Graeme Wilson
Richard Wilson
Rupert Wilson
Peter Winfieldale
Mike Winter
Niels Winters
Stuart Witts
Ian Wolloff
Brodie Wolstenholme

James Wolters
Lee Woodberry
Damian Woodings
John Woods
Julian Woods
Steve Woodward
Douglas Wooldridge
Anthony Woolhouse
Benjamin Woschek
Brent Wouda
Matthew Would
Tomasz Woźny
Dave Wright
Elaine Wright
John R Wright
Michael Wright
Michael Wulff Nielsen
Simon Yapp
Alan Yarker
Bailey Yates
Dave Yates
Sam Yates
Kent Yee
Slavi Yordanov
Peter Young
Steven Yu
Leo Zalewski
Maxim Zapryanov
Dominic Zartarian
Marc Zeguers
Lennart Zeidler
Julie Zeraschi
zerocarbon.vision
Nikolai Zheleznikov